Gaia's Body

Gaia's Body

Toward a Physiology of Earth

TYLER VOLK

COPERNICUS

AN IMPRINT OF SPRINGER-VERLAG

© 1998 Springer-Verlag New York, Inc.

Published in the United States by Copernicus,
an imprint of Springer-Verlag New York, Inc.

Copernicus
Springer-Verlag New York, Inc.
175 Fifth Avenue
New York, NY 10010

Library of Congress Cataloging-in-Publication Data
Volk, Tyler.
    Gaia's body : toward a physiology of earth / Tyler Volk.
        p.   cm.
    Includes bibliographical references (p.   ) and index.
    ISBN 0-387-98270-1 (hardcover)
    1. Gaia hypothesis.   I. Title.
QH331.V714   1997
577'.01 — DC21   97-24365

Manufactured in the United States of America.
Printed on acid-free paper.
Designed by Irmgard Lochner.
Preface photograph by Lyn Hughes.

9 8 7 6 5 4 3 2 1

ISBN 0-387-98270-1   SPIN 10557106

1/00

Contents

Preface: Fantastic Voyagers

Before leaving New York for the year to write this book, I sought out an old film. Packing and other preparations for the move to New Mexico had grown frantic. But I was driven to take the time to relive an adventure story I had seen as a teenager, with the hope of gleaning inspiration for this book. So, a few days before jetting west, I rented *Fantastic Voyage*.

The hero in the story is a government trouble-shooter. One night he is rushed to a briefing deep within a secret underground laboratory. An anxious clamor surrounds the body of an unconscious diplomat who has been critically injured by foreign agents. Fortunately, the laboratory possesses an astounding new invention, an all-purpose miniaturizer. The plan is to put the hero and four others—a driver, a helper, a surgeon, and Raquel Welch—inside a submarine, shrink all to the size of a blood cell, and then launch them into the diplomat's body. Once inside the blood stream, the crew will attempt to navigate the sub to the brain and there, using a laser, clear an otherwise inoperable blood clot.

The briefing ends. The team is asked, "Is there anything else you need?" The hero, who had been growing ever more dubious, says, "Yeah, a taxi to get me outta here." Too late for that, buddy. On the other hand, the surgeon's assistant, played by Ms. Welch, can hardly wait to climb aboard.

In their nanosubmarine, the nanovoyagers are injected by hypodermic needle into an artery in the diplomat's neck, headed for his head. Blood cells gyrate around them in an acid-rock light show (the film is vintage 1966). They stare slackjawed, like ecotourists on the Amazon River.

Things do not go as planned. A hair-raising detour forces them to navigate through the heart—temporarily halted for their safety. Eventually the intrepid explorers enter the left lung. The blood passageway narrows and narrows. Undulating blobs magically shimmy from blue to red. The voyagers have become the first in history to witness the revitalization of blood cells, as those cells release carbon dioxide and absorb oxygen across a supply capillary's thin membrane. Because the submarine is low on oxygen, the hero dons scuba gear, swims over to the membrane, and snakes an airhose across it.

Why, after more than thirty years, was I so keen on seeing this movie again? The idea of traveling inside a giant body resonates in my mind like an archetype. I am a voyager inside the biosphere.

We need not wait to be shrunk by some futuristic wizardry to witness an exchange of gases from an interior viewpoint. Just walk in the woods. Trees are taking in carbon dioxide and releasing oxygen across the membranes of their leaves. Each of us already has an airhose, the trachea, by which we tap into the atmosphere and inhale what the trees freely offer. Take a deep breath. Or swim amid the living colors of a reef. Lie down in a field of undulating grass. Let a hand bob in a river's eddies. Smell some pungent dirt. We are already fantastic voyagers. All of us are cells within the embracing physiology of what Jim Lovelock has called "Gaia."

Thinking of Earth's life-inhabited surface as a *physiological* system immediately conjures up an image of a giant volitional being, as does naming the system after a Greek goddess. We thus must exercise some care in applying this analogy. Organisms evolved; Gaia did not. Beyond alluding to such concepts, I will not weave detailed arguments for or against the idea that Earth is alive, that Gaia self-regulates, and that

Gaia is a self-sustaining organism, or perhaps quasi-organism. Such notions depend on a slew of ambiguous words that, however carefully defined, either ready readers for an Earth-hug or raise their hackles. In either case, the reader's attention to the *science* of Gaia and its overarching principles may lapse. I will therefore try to work with a tricky binary. On the one hand, I experience a delightful sense of being inside a giant metabolism. This perception grows more acute the more I learn, but I am also convinced that Gaia is very different from any organism. Thus I can honestly apply the principles of science to study the global metabolism without postulating a global organism.

What is Gaia? Following Lovelock, I consider Gaia the interacting system of life, soil, atmosphere, and ocean. It is the largest level in the nesting of parts within wholes that encompasses—and thus transcends—living beings, a nesting that ranges from the molecules within cells all the way outward to the gaian system itself. Like the interiors of organisms, Gaia contains complex cycles and material transformations driven by biological energy. Indeed, Gaia's inclusion of life means that from some perspectives, it much resembles life. But how Gaia differs from organisms turns out to be its glory.

Consider: Although Gaia has changed through time, it does not evolve in a Darwinian sense. Nevertheless, it both contains and is built from evolving organisms. Furthermore, organisms are open, flow-through systems, whereas Gaia is relatively closed to material transfer across its borders. Gaia exists on its own unique level of operating rules, a level surely as complex as that of organisms and therefore worthy of its own science—which Jim Lovelock calls geophysiology.

Gaia is an entity whose properties we are just beginning to understand. How can that be? Oxygen and carbon dioxide, after all, have been known as a functional pair since 1779, when Jan Ingen-housz found that plants made "vital" air that would sustain animal life, during the day, and "vitiated" air, dangerous to animals, at night. The scientific question for our own time is why the "vital" component of air holds at 21 percent. On such a fundamental mystery alone I could rest my case that our voyage of discovery has barely begun. But consider further:

Only since 1958 have measurements of carbon dioxide enabled us to witness the seasonal breathing of the biosphere. And only recently has it become possible to monitor the riverine transport of dissolved ions and particles to the ocean, revealing what flows from land to life at sea. Daily we hear announced the discovery of new microbes deep in rock, or underwater, or even in backyard soil. Data for the big picture—the gaian picture—are just now pouring in.

For many years my professional work has been involved with Earth's carbon cycle, with crop growth for NASA's closed systems, and with evolutionary innovations that affected climate. Throughout, the focus has been on the role life plays in chemical cycles and on the complex interaction between life and the environment. This research led me naturally to the Gaia hypothesis of Jim Lovelock and Lynn Margulis, and I consider myself privileged to have personal and professional friendships with many scientists and nonscientists alike in what might be called the gaian community. We are all intrigued by the magnificent mysteries that loom before us as we ponder how this planet works.

It is from such a background that I offer this book. I believe that new insights will derive from contemplating the long known and newly known from fresh angles and that such re-visioning will invariably take us to the edge of the unknown. In articulating my personal vision of Gaia as a symphony of material flows and cycles, I have neglected some classic topics in the gaian literature, such as the Daisyworld model of climate regulation, debates about the theory of Gaia, and the history of gaian ideas and research. Readers will find these topics beautifully presented in Lovelock's own books and in the writings of others.

My focus is on the molecular transformations between life and the global environment. Atmospheric carbon dioxide, for example, seasonally rises and falls with the shifting dominance of respiration and photosynthesis in the Northern Hemisphere. This "breathing of the biosphere" leads, in Chapter 1, to a number of directives for studying Gaia: We should follow the cycles of matter. We should attend to the cycles of causes, in other words—to the feedback loops of influence that control the cycles of matter. We should look for key fluxes and agents. And in accordance with the prime directive for gaian inquiry, we should look to the profound difference that life works upon a planetary surface.

The second chapter, "A Global Holarchy," develops some conceptual underpinning for treating Gaia as a whole made up of parts. Parts influence their wholes by outward causation. Our exploration starts among the denitrifying microbes in the ocean. Then we follow the released molecules as they travel upward into the atmosphere and downward again by the efforts of nitrogen-fixing bacteria. Wholes, in turn, influence their parts by inward causation. How does Gaia influence its parts? I suggest that the fundamental effect results from the huge difference in magnitude between the flows and cycles rumbling inside Gaia and the relatively small leaks of molecules across Gaia's borders. Gaia's border frames the creativity of life within.

The next three chapters trace a path through the holarchy from large to small. "Outer Light, Inner Fire" examines the consequences for Gaia of the immense presences above and below, whimsically called Helios and Vulcan. Sunlight falling as parallel rays on a spherical surface endows various regions unequally with energy, and this inequality drives the whirls of atmosphere and ocean that freely provide Gaia with a circulatory system uniting life. Below Gaia churns the deep Earth with its heat. A minor source of raw energy, Vulcan nonetheless powers key transfers of material across gaian borders at ocean ridges, through volcanoes, and by subduction.

Chapter 4 moves from these outer agents to the "parts of Gaia." What are its parts? Are they perhaps the biomes? The genetic divisions of kingdoms or domains? The cycles of elements? All these viewpoints combined? One perspective focuses on the biochemical guilds: groupings of organisms that perform similar chemical functions (photosynthesis, for instance). Such guilds may be the closest analog in gaian physiology to living organs. Another potent approach to Gaia's parts emphasizes the four primary pools: life, soil, atmosphere, and ocean. Life, a minor pool in terms of its store of crucial elements, exhibits its strength as a major player in the Earth system by generating surface areas of leaves, algae, fungi, and bacteria that sum to the nearly astronomical in size.

The living surfaces only hint at how organisms are interlaced with the gaian matrixes of soil, atmosphere, and ocean. Chapter 5 continues the inward journey, to the little molecules that run the world, the "worldwide metabolisms." For example, the green chlorophyll of plants, algae, and cyanobacteria unites these disparate kingdoms into a single biochemical guild of water-splitting photosynthesizers. Another universal is the enzyme Rubisco, which ushers carbon dioxide into the biological molecules of plants and algae. More worldwide molecules directing the cycles of nitrogen and phosphorus indicate that, overall, the key processes of Gaia are best seen by looking right through the borders of organisms.

All these fluxes of matter between life and the gaian matrixes are driven by energy. How much of the sun's energy is bound into biochemical energy by photosynthesis? When we apply an analysis called the energy cascade, the theoretically possible number turns out to be surprisingly small, and the real number is only a tenth of that. How, then, can life have any impact on the planet? The answer, explored in Chapter 6, is in how that "embodied energy" is used: to drive chemical transformations that would be either absent or rare without life—for example, mining water molecules for precious hydrogen and thus freeing oxygen as waste. Further, life embodies not just energy but also a mix of elements that is amazingly uniform in proportions across the organic spectrum. The feeding of life on its own productions amplifies the availability of crucial elements such as terrestrial phosphorus to high values, a boost that life gives to itself by tightening the cycles of matter within Gaia.

Chapter 7, "The Music of This Sphere," continues the theme of the cycling ratios after elements travel down river to the sea. There, for example, the cycling ratio of phosphorus is even higher than on land. There too we encounter a duet between phosphorus and nitrogen played on the grandest of scales: in their steady ratios as dissolved nutrients at all depths of the global ocean. What brings about this chemical regulation? For an answer, we revisit the microbial activities of denitrifiers and nitrogen fixers. Other interactions include iron, sulfur, and then clouds, climate, atmospheric carbon dioxide, and the biotic enhancement of weathering.

Finally, in Chapter 8, we contemplate "Gaia in time." Gaia's story involves the emergence of many of the biochemical guilds in life's earliest days (geologically speaking). But there was plenty of room left for the evolution of new guilds between that distant then and now, as the environment altered by previous forms of life opened opportunities for new guilds. Gaia owes its continuity during its long tenure to the nearly closed network of life and life's wastes, both organic and inorganic, held within the gaian matrixes. This is the physiology of Earth. How, in the end, should we regard ourselves in this story?

The crew in *Fantastic Voyage* do reach their goal. They blast away the blood clot; the diplomat survives. Moreover, my idiosyncratic craving to watch this film during the domestic madness of a major move was well rewarded. "Fantastic voyage" is an apt metaphor for the scientific quest to understand Gaia. To repair damage caused by humans, the crew travels through the passageways of a body made by nature. Isn't that our current situation? We inhabit a global metabolism with a four-billion-year pedigree. In just the past few decades, we have awakened to an awareness of damage we ourselves are inflicting on this metabolism with our blind urges to procreate and appropriate. Perhaps in the fantasies of many, the "taxi outta here" would be a time machine. Just tell the time taxi to stop at nature a thousand or more years ago. But because most of us would want to keep the postal service, the Internet, MRI scans, and an abundance of items in stock at a local supermarket, we must proceed with the world as it is, and that requires knowing how the foundational processes of nature work. I am convinced that such knowledge, if widely held, will contribute to shaping the future of

Gaia—a future in which we, as a new biochemical guild, will necessarily be integrated into the global metabolism, for better or for worse. We can make it for the better by promoting an informed reverence for Gaia's body.

That said, let the journey begin.

Acknowledgments

For help in this project in a variety of ways, I thank Bob Berner, Wade Berry, Richard Betts, Bruce Bugbee, Ken Caldeira, Donald Canfield, Jim Cavazzoni, Orren Champer, Paul Falkowski, Victor Gallardo, Richard Geider, Terri Gregory, Kevin Harrison, John Hedges, Mae-Wan Ho, Dick Holland, Lyn Hughes, Markus Hüttel, Sonja Jones, Jim Kasting, Dave Keeling, J. Gijs Keunen, Lee Klinger, Lee Kump, Rocco Mancinelli, Lynn Margulis, Dan McShea, Euan Nisbet, Hans Paerl, Michael Rampino, John Richards, Heide Schulz, Mark Sutton, Francesco Tubiello, Ben Volk, Bess Ward, Mike Whitfield, and Tim Whorf.

I am grateful to Marty Hoffert, Tim Lenton, David Schwartzman, and Ron Williams for reading and commenting on substantial sections of the manuscript. Special appreciation goes to Jim and Sandy Lovelock for their invitations to the Oxford Gaia conferences in 1994 and 1996, and for their warmth and inspiration over the years. For wilderness hospitality and friendship, many thanks to Fred Norton.

New York University provided an intellectual home for thinking about the whole Earth and granted me a half-year sabbatical and half-year leave to work on this project. The excellent services of the Interlibrary Loan Department of the Miller Library at Western New Mexico University proved invaluable. The Space Science and Engineering Center at the University of Wisconsin produced the satellite images from

NOAA that appear throughout. Lyn Hughes kindly provided the photograph on page xiv. (Other sources of images are acknowledged in the figure captions; unacknowledged figures are by the author.)

I am obliged to the professional efforts of many at Copernicus Books, including supervising production editor Steve Pisano, assistant editor Teresa Shields, and copy editor Connie Day, who refined the entire manuscript and delighted me with suggestions of alternative wording. Bill Frucht, senior editor and resident visionary at Copernicus, put the buzz in my ear about a book on Gaia. I thank him for wise and thoughtful editorship along every step of the way.

Finally, I thank Connie Barlow for many insights through the years about Gaia that I probably now think were mine, and for giving her exceptional touch to all aspects of this project.

Gaia's Body

1
Breathing
of the Biosphere

S ilver City, New Mexico: Into the cloudless, burning blue skies
of late June, just when the desert landscape is starting to bake
from months of drought, fingers of moisture draw up out of the
Gulf of Mexico. Looking north, toward Colorado, I imagine the air
rising across the entire broad plateau of the American southwest, sent
skyward by a combination of sun and elevation. In the wake of the lifting
continental air, marine air from the south is inhaled landward from over
the bathtub-warm Gulf. For a week or two the heralds arrive in waves:
white clouds, often accompanied by lightning and thunder and some
promise of gray, but precious few drops of rain. All the hills and canyons
are thirsty and waiting. The monsoon is coming.

By mid-July the skies are exploding daily. Crystal clear dawns often lead to dark, swirling torrents, even hail. Sometimes the sun returns later in the afternoon. At other times the storm fronts a vast sweep of overcast, which may stay as long as a day, bringing the best gift of all: a steady drizzle.

As I watch the weeks of transition and think about all the forces that coalesce to create the change of state, it seems as though the pounding heat of June can last only so long before something in the system shifts. "Gaia can tolerate just so much sun before cooling off," I say, watching the clouds almost miraculously appear.

Why would I, a reputable scientist, invoke a name out of Greek mythology? Gaia was the ancient Greek goddess of the Earth—in fact very ancient, for Gaia is older than Zeus and Athena and most of the other familiar gods and goddesses. But I should need no goddess to explain what's happening, because I am versed in the basic atmospheric dynamics of the monsoon. If the Old Testament Lord as voice from the whirlwind had asked me, as he asked his earthly servant Job, about who fashions the clouds and brings the rains to sweeten the desert, I would not be struck dumb as Job was by ignorance about the elementary workings of the immense and nurturing world. I would repeat the facts of the monsoon I outlined just now.

What if the Lord (a.k.a. Yahweh, a.k.a. Allah) were then to challenge me further: "OK, Dr. Know-it-all, how will the clouds respond and thus affect the global warming projected from your rising levels of that greenhouse gas, carbon dioxide?"

"Uh, the best atmospheric scientists of my generation all admit that clouds inject the most uncertainty into climate prediction, accounting for a difference of a factor of three in the projected warming. So I certainly don't know."

And what if the whirlwind were to continue pushing. "OK, Dr. Doesn't-quite-know-it-all, from the retrieval of ancient air locked in the deeply buried ice sheets of Greenland and Antarctica, your colleagues have discovered that before you all began burning fossil fuels, the carbon dioxide was even lower during the Ice Age, say twenty thousand years ago. What explains that lower amount?"

"Well, some truly insightful theories abound, having to do with an altered ratio of the bodies of plankton to their shells in ocean detritus, or a lowering of overall ocean temperature, or a change in chemistry of the polar seas. These theories all explain some pieces of data from that icy time, but the puzzle has not really been fully put together. Again, I don't know."

The whirlwind might spin some more conundrums my way: "Then let's take something simple, Dr. Knowing-less-and-less-by-the-minute, something every school child knows. That vital gas, oxygen, which makes up 21 percent of the total atmosphere—you do breathe it, don't you?—surely you must understand the workings of the Earth that have made it that 21 percent?"

"Ah, yes. We know it is derived from photosynthesis, when the products—parts of marine algae primarily, but also tree bark and other tough, organic substances—get buried and locked away into rocks. But why 21 percent? Possibly something to do with how much phosphorus gets recycled into the water during the ocean burial of those dead algae, but this is only a recent, tantalizing clue. I'm ignorant again."

Maybe I would finally wise up during one of the rounds in this ever more humbling interrogation and just say, "Gaia made it so" or "I don't know, go ask Gaia." Then perhaps the Lord would want to meet this Gaia and I'd be off the hook.

But this strategy would not be true to my use of the concept of Gaia. I don't use Gaia as a cop-out to mask my ignorance, as an explanation for the currently unexplainable. Gaia for me is an intermeshing between the known and the unknown, inseparable. Still, why Gaia? Why have I enthusiastically written technical papers about Gaia? Why have I jumped at the chance to attend all major scientific conferences about Gaia?

For a short answer, let me say that Gaia has been and continues to be the best reminder—a call directly out of science itself—of something stupendous to research, to ponder, to use as a source of ideas. That something has much to do with the Earth as seen as a whole from space. That vantage point reveals a sphere, glistening before our eyes—

a unique, integrated system of white gyres, deep blue oceans, a wispy blanket of gas, and rocky brown continents blanketed with the green of life. I use Gaia as a reminder to think "global" and to think "system."

Historically, the name Gaia (Gī'–ah) was bestowed on this immense entity in the 1970s by the eminent British scientist James Lovelock. His concept, in a nutshell (from a 1996 writing), points to

> a tightly coupled system whose constituents are the biota, the atmosphere, the oceans and the surface rock . . . [with] self-regulation of important properties like climate and chemical composition . . .

This certainly doesn't sound much like a goddess. It's not clear that the thing described is even alive. But perhaps that depends on what one is looking for in one's deities; perhaps it depends on what one calls alive.

The gaian system includes the totality of life, from the tiniest squiggling bacteria miles deep in the ocean to the high-flying sandhill cranes overtopping Mount McKinley in Alaska. It also includes all that life directly touches and affects. The gaian system thus extends above the cranes into the stratosphere, because ozone there derives from oxygen, and oxygen from life. It extends into the sediments of the deep ocean, where benthic bacteria and other creatures living on the rain of debris from the ocean's surface affect the chemistry of the water. The gaian system, then, constitutes an immense whole; furthermore, we are just beginning to understand its organizational properties.

I consider Lovelock's naming of this special system a wake-up call to science, a reminder to think about special properties of the whole. Certainly, he and his collaborators, such as microbiologist Lynn Margulis and marine chemist Andrew Watson, have not been alone in sounding the clarion call to "think global." NASA, for example, has sponsored research under such umbrella titles as Global Ecology, Biospherics, and (most recent and renowned) Earth System Science. A set of advanced satellites for integrating the study of everything from caribou populations to ocean gyres is nearing launch. Called the Earth

Observing System, it is better known as EOS, formerly the Greek goddess of dawn. NASA, too, seems fond of goddesses.

We all seem to be thinking global these days. But Jim Lovelock's personal and impassioned tenacity in using the name Gaia in his scientific papers, books, lectures, and conferences for over twenty years, as well as the efforts of others who have been inspired to use the word, could be taken as an exceptionally apt way of saying thanks to something bigger, even better, than a mere human self. Contemplation of that something has animated ideas and discoveries and has spawned ideas for future discoveries. So I honor the word Gaia. It personalizes our relationship with the planet, an attitude we (and the planet) could use a lot more of.

William Blake once wrote that gratitude was heaven itself. Life, to be sure, is essential to gratitude, life that as a totality preceded our new-sprung human kindling with a steady global fire of green for nearly four billion years. But perhaps paradoxically, the seeming absence of life in the dynamics of the monsoon gives special meaning to my invocation of Gaia in that context. It helps me remember that Gaia is not just a pet name for the sum of living beings. Rather, it includes air, water, and even clouds. Though life will occupy center stage in much ⵑ my discussions of Gaia, the other parts, too, deserve a share of the spotlight. (Can our language do no better than name them by negation: inanimate, nonliving?)

Clouds are complex enough. But throw life into any situation and the plot of nature turns from a nursery rhyme into a Russian novel. Life as an active force is what makes the puzzles of the reduced carbon dioxide during the Ice Age and the stability of oxygen at 21 percent so enigmatic. What is the global role (or roles) of life? What happens when you factor life in as part of the system?

If that question could be answered in a sentence, I would not be writing this book.

What about life in the monsoon? Plants secure the soil, which holds the water that the plants draw upward through their roots, transport across their leaves, and release into the sky. This transpiration of water vapor can help feed the next round of rain. Furthermore, the monsoon's strength is definitely tied to the general state of Earth's climate, and thus to carbon dioxide, the carbon cycle, and carbon-based life, including us. How will today's rising level of carbon dioxide alter the monsoon? For it is certain that it will.

To begin thinking about Gaia, we need not ponder the extinction of the dinosaurs, or think about how the sun has changed over four billion years, or contemplate New York City's lot under thick ice just a geological fingersnap ago. We can look at what's happening right now. For example, in the global carbon cycle, intensely investigated yet still mysterious, we can witness the physiology of the planet. With the carbon cycle, we can begin to wrap our minds around the entire Earth.

ᔓ THE BREATH OF GLOBAL LIFE

We commonly note famous people of the twentieth century, famous events, and even famous movies and books. But rarely do we sit up and take notice of graphs.

I've got a candidate for the all-time list of famous graphs. Depending on what happens to climate in the twenty-first century, it may end up being number one on the list. In my ranking and those of many others, it's already near the top.

This remarkable graph shows the increasing amount of the most notorious greenhouse gas. Carbon dioxide — CO_2 — acts like a one-way valve to energy. It lets the sun's energy pass through, mostly as short-wave, visible light, right down to the surface. But with regard to Earth's own energy headed upward from the surface toward space in the form of long-wave, infrared rays (invisible to our eyes), the valve in the atmosphere shuts. Carbon dioxide blocks infrared rays by absorbing them. It then re-radiates about half of them back toward Earth's surface. As more CO_2 pours out of smokestacks and exhaust pipes into the sky from the burning of carbon-based fossil fuels — largely the big three of oil, coal, and natural gas — more of Earth's infrared rays are blocked, and more of the surface's energy rebounds from the sky, which ultimately makes that surface warmer. The resulting extra warmth is called the greenhouse effect. Exactly how much extra warmth we get will determine the ranking of the CO_2 graph decades from now (the warming could be predicted better, if it weren't for those clouds).

A small amount of carbon dioxide is causing all this concern. Nitrogen, for example, which makes up some 78 percent of air, is more than two thousand times as abundant. Oxygen is more than five hundred times as abundant. But neither of these is a greenhouse gas because each has only two atoms in its molecules—too simple a geometry to absorb and re-emit the infrared frequencies. So scanty are the molecules of carbon dioxide that its concentration is cited in the scientific community not as a clumsy fraction of a percent (0.036 percent), but as the number of molecules in a random sample of a million molecules of air—in other words, as parts per million. The concentration of CO_2 is currently nearing 365 parts per million, or 365 ppm, and rising.

This bit of background sets the stage for the second most important feature of the CO_2 curve, after the fact of its inexorable rise. The preindustrial value since the final collapse of the great ice sheets was about 270 ppm, which we know from analyzing bubbles of ancient air preserved deep in glacial ice. Only in 1958 did real-time measurements begin, thanks to the scientific mettle of Charles David Keeling. By then the carbon dioxide concentration had already escalated to 315 ppm. Since then, daily measurements at the lab site Keeling founded have provided a wealth of fascinating data. Let's zoom in on a few years of the "Keeling curve" generated at the lab atop Hawaii's lofty volcano, Mauna Loa, in the central Pacific Ocean.

Superimposed on the overall upward trend are annual cycles strikingly visible on the graph. Unexpected at the time of discovery, these cycles have been much scrutinized. Some process or combination of processes subtracts CO_2 from the atmosphere and then faithfully adds it, year after year. At Mauna Loa, the low occurs in September or October, dipping between 3 and 4 ppm below the average upward trend. The value peaks in May, between 3 and 4 ppm above the trend. Because trend and cycle are combined, the measured rise from October to May

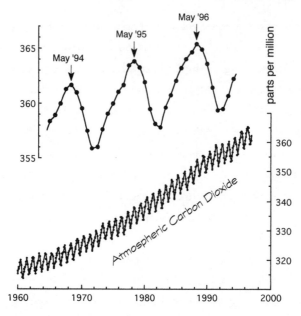

Carbon dioxide at Mauna Loa, Hawaii. Within the upward trend caused by fossil fuel combustion, the biosphere's breaths rise and fall with the seasons. Data for this site (20°N, 156°W) provided by C. D. Keeling and T. P. Whorf at the Scripps Institution of Oceanography and by the Carbon Dioxide Information Analysis Center at the Oak Ridge National Laboratory.

is greater than the measured drop from May to October. But considered by itself, the cycle currently goes 7 ppm up and then 7 ppm down.

The cause of this oscillation? Not a seasonal cycle of volcanic emissions from Mauna Loa. Not a seasonal cycle of tourists. The source of the cycle is not, in fact, local. Imagine how quickly winds mix the atmosphere, how weather that was a thousand kilometers away yesterday arrives here today. Mauna Loa's remoteness, in the central Pacific at the northern edge of the tropics, has made it a perfect site for sampling the average air of the entire Northern Hemisphere.

An obvious suspect for producing the cycles of carbon dioxide would be fossil fuels, burned mostly in the Northern Hemisphere. Wintertime heating of our habitats could impart a seasonality to the release of the greenhouse gas, and it would act in the right direction at the right time. But it turns out that cool-season heating is pretty well compensated for by summertime air conditioning and increased pleasure driving. Experts who compile the carbon budgets have found that the seasonal variation in fossil fuel release is surprisingly slight. So nix fossil fuels.

The ocean, an immense and complex part of the carbon system, is another suspect. Guessing that the ocean is behind the atmospheric oscillation leads to a simple test. What is the cycle in the Southern Hemisphere at the latitude equivalent to Mauna Loa's? Ships that have measured carbon dioxide at 20° south latitude report a surprise: almost no atmospheric oscillation. This might seem odd, because temperature and biological activity do cause the pressure of carbon dioxide to vary in the water itself. Why the ocean exerts so little influence on the atmosphere owes to a combination of factors, including the relatively slow rate of gas transfer. But we do not need all the theoretical underpinnings. The Southern Hemisphere data tell us what we needed to know: nix the oceans.

Who's left in the line-up of suspects? At conferences Dave Keeling often wished aloud that the air's CO_2 molecules were color-coded according to their origins. Those from fossil fuels would be stained gray. Those from the ocean would be dyed blue. And those that had passed though land plants would of course be painted green. Rule out gray and blue, and green remains. If molecules could be painted, the curious annual cycle in the air at Mauna Loa would be visible as an alternating increase and decrease of green molecules—a global breathing.

Gaia founder Lovelock and German geochemist Wolfgang Krumbein have been employing the word *geophysiology*. Certainly the Mauna Loa cycle shows geophysiology in action, the alternating dominance of photosynthesizers and respirers on the grandest of scales. When building molecules from the energy of absorbed light, the photosynthesizers inspire CO_2 and, except for some types of bacteria, expire O_2. Included are trees both broad-leafed and needled, slender grasses, delicate mosses, the algae of lichens, tiny marine diatoms in glassy shells, coccolithophorids clothed in calcite, underwater towers of kelp off rocky seashores, symbiotic algae living within the bodies of coral, and (last but not least) the ubiquitous cyanobacteria, blue-green with chlorophyll, in soils, abounding in fresh waters, and adrift in the seas. The photosynthesizers include everything green with chlorophyll and some types of algae and microbes colored purple, red, or brown by other dominant pigments.

The respirers include all organisms that inspire O_2 and expire CO_2: aerobic bacteria toiling as waste processors in the soils and within the ocean, from its surface to the dark benthos miles below; fungi toiling as waste processors and nutrient grabbers in the soils and attached to plant roots; single-celled denizens of the oceans equipped with filaments of protoplasm for capturing prey, such as formaniferans and radiolarians; all the animals of land and sea, including yours truly who consumes a pound or two of oxygen per day and produces about the same in carbon dioxide; and (last but not least) the photosynthesizers when they are not photosynthesizing.

The global breath often rushes in both directions simultaneously. Consider the grape vine beyond my porch, which drapes over the cedar tree's droopy limbs. It's five P.M., the sun is perfect, and both tree and vine are actively absorbing carbon dioxide and releasing oxygen. But in the soil beneath them, the bacteria, fungi, and even cells of their own roots are releasing carbon dioxide and consuming oxygen from the fresh air that seeps into the soil. The carbon dioxide filters up, rising out of the soil and into the air. Thus the system of life as a whole is both absorbing and releasing both carbon dioxide and oxygen. If these two opposing processes were always everywhere in balance, we would measure no oscillations in the record at Mauna Loa. But they're not.

Where the temperature varies little year-round, mature tropical forests and their soils, as ecosystems, may maintain a near balance at all seasons, but at most tropical sites, wet and dry seasons alternate and set up a rhythm of gases. The further poleward one treks, the more the gases pulse with the thermal seasons. During late spring and summer, the inhalation of carbon dioxide dominates the net breath across temperate and high-latitude ecosystems. Note that the key concept rests upon the *net breath* as the sum of ecosystem gas fluxes. Because the heat drives their bacterial inhabitants to a frenzy, soils in the temperate and high latitudes actually exhale more CO_2 in summer than in winter. But to an even greater degree, the inhalation of green plants also crests then. In fall and winter, photosynthesis plummets to near zero. During those seasons, respiration, though reduced from its warm-weather heyday, continues uncontested by photosynthesis and thus truly gets its day.

Who could have known that this annual breathing of the land would seesaw the carbon dioxide over the middle of the Pacific Ocean? But there it swings: 7 ppm down in spring and summer, 7 ppm up in fall and winter. That's about two percent of the air's content of CO_2 for the entire Northern Hemisphere. And that's a lot of variation in the chief gaseous feedstock for chlorophyll-carrying members of the photosynthesis club.

The cycle at Mauna Loa reflects the net flux across continental ecosystems—the difference between photosynthesis and respiration. Thus the total carbon incorporated into plants during late spring and summer is substantially more than the net, even in places where ecosystems are highly seasonal. To work an example, allow that the total warm-season uptake by photosynthesis is three (arbitrary) units. (The plants' own respiration at night has already been folded into their three units.) Simultaneously, the respiratory release of CO_2, primarily from soil bacteria, is about two of the same units. Taking the difference between these opposite fluxes yields a net ecosystem flux equal to one unit. I've set these numbers in about the right ratios for latitudes above $30°$ north; the flux into plants typically exceeds by about three times the net influx for the ecosystem itself during the season of growth.

In the lower latitudes the seasonal cycle is less pronounced, so respiration by soil organisms is an even larger fraction of photosynthesis during the season of maximum plant growth. If all photosynthesis globally is accounted for, it annually bites out about ten percent of the atmosphere's carbon dioxide. The two percent seesawing in the Northern Hemisphere data at Mauna Loa is like the tip of an iceberg compared to the activities of the much heftier players on the seesaw itself. Imagine if respiration, for example, were to step off the seesaw and all plants

11

could continue their normal cycle of photosynthesis. Carbon dioxide would drop rapidly, so rapidly that the atmosphere would be emptied of CO_2 in about ten years. We all like to feel part of the world, but perhaps being labeled respirers doesn't do it — sounds like we are leeches on Gaia, rip-off artists. Yet clearly the plants would starve without the respirers' exhalation. Breathe out deeply. Hooray for us. But the bacteria deserve the most credit.

In the event that carbon dioxide levels should plummet, the nearly forgotten realm of Poseidon would come to the rescue. Its immense store holds fifty times more carbon dioxide than in the atmosphere, and the ocean, like root beer, would outgas. That factor of fifty lengthens the ten-year supply to about five hundred years. The reprieve is only temporary. Even several thousand years would be precious little room for safety. As the cycle of the blood sustains our bodies, so the cycle of the complementary breaths of photosynthesis and respiration is necessary for the continuity of the biosphere.

Concern with the fate of our emissions of carbon dioxide from the burning of fossil fuels led to the establishment of additional labs, after the initial site at Mauna Loa. The second was founded at the South Pole. By the 1980s a global network was up and running. We can now monitor the biosphere's breath in detail at different latitudes. With a graph in three dimensions — CO_2, time, and latitude — we can witness the breathing of the biosphere in even richer glory.

This graph in 3-D shows that the Southern Hemisphere has almost no seasonal cycle of carbon dioxide. The cycle is driven by temperate and high-latitude ecosystems, but Africa and South America taper off at these crucial latitudes, and the only substantial land, Antarctica, is locked in ice. In contrast, the continents of the north balloon in high latitudes (think of the boreal expanses of Canada and Russia), which also is reflected in the 3-D portrait. The higher the latitude in the Northern Hemisphere, the more extreme the seasonality of photosynthesis and respiration, the more vigorous the amplitude of the CO_2 cycle. At Barrow, Alaska, the seasonal breath swings up and down by more than twice the average for the hemisphere at Mauna Loa.

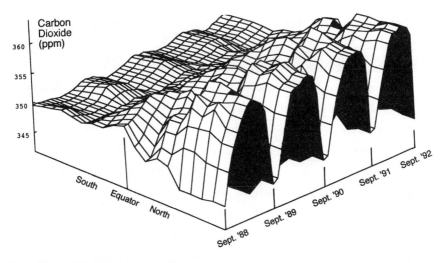

Breathing of the biosphere in three dimensions. This graph of atmospheric carbon dioxide, with monthly time intervals and 10° latitude intervals, was constructed from fifteen sites, ranging from the South Pole to 82°N in Canada. Data provided by the Carbon Dioxide Information Analysis Center at the Oak Ridge National Laboratory.

The three-dimensional breathing of the biosphere reveals much about geophysiology: the difference in the fractions of land and ocean between the hemispheres, the binary cycle of photosynthesis and respiration, the unity of life with the atmosphere. This graph belongs on T-shirts in the streets of Honolulu, New York, St. Petersburg, Paris, Tokyo, Rio.

∽ CYCLES OF CAUSES

About the coming of the monsoon I could have simply said, "The clouds are making it cool." When clouds shield the sun, my skin goes from toasted to comfortable, this is true. But the agents are more than the clouds. A web of causes creates the monsoonal rains: the seasonally high sun, the elevation of the southwestern plateau, the giant bathtub to the southeast and the water vapor it supplies, even the physics of convection, whereby heat within the thunderheads released during condensation at low levels pumps the swirls up to still higher altitudes, releasing more heat in a leap-frogging process that can end in a pelting of hail on what had been parched days. Saying that it's just clouds that are making it cool limits our thoughts to a sole cause, which restricts our view of a complex system to a linear cause-and-effect relationship. (This sort of oversimplification is not unknown, for example, in the political and economic analyses of television land.)

I know of no way to head more directly into Gaia's web of causes than the study of carbon. Thinking about nitrogen, phosphorus, oxygen, or sulfur, the mind also traces great natural cycles. But carbon is our primary focus because of the greenhouse experiment we are conducting. Scientific meetings on the carbon cycle have been wonders to behold: plankton ecologists talking late into the night with agronomists, and both with climatologists and chemical oceanographers (OK, that happens mostly during the day). The sessions leave no stone unturned; geologists arrive with isotope data that suggest values for carbon dioxide a hundred million years ago. The humble carbon atom has been teaching earth scientists how to combine the local research that characterizes their disciplines into a new global science. Just as the science of physiology began with understanding the cycle of blood, future historians of science may date the origins of geophysiology from the first steps in understanding the carbon cycle.

Perhaps the entire carbon cycle should be on T-shirts. What would it look like? Would it fit?

A carbon atom is like a mountain lion in that it takes its rests in a complex sequence of places. In a *ménage à trois* with two atoms of oxygen, it may waft in the air above the seas and continents for years before entering a stomate, or pore, of a grass leaf in the pampas of Argentina. Immediately forged with hydrogen into a simple sugar, by midnight the carbon atom, burned and its energy sacrificed, may be respired again in airborne CO_2. Alternatively, in reactions fueled by the sacrifice of other carbon atoms, it is about equally likely to remain longer in the plant, being synthesized into cellulose in the stalk, new chlorophyll in the leaf, or lipid in a border cell of a root. If it indeed winds up within a border cell, it soon is sloughed off as the growing root rubs against soil particles, whereupon it is probably digested by a root-associated bacterium. Thus oxidized into carbon dioxide, and now a part of the air trapped in soil, it diffuses upward into the atmosphere to float above the seas and continents for years again.

While still in the soil's interstices of air, however, the respired molecule of carbon dioxide could have taken another path altogether. The odds of its taking this path are about one in a hundred. Traveling in groundwater with calcium ions, the carbon—now in the form of a bicarbonate ion—eventually reaches the Paraná River and ultimately the southern Atlantic Ocean. Even if it had taken the path of the other 99 and re-entered the atmosphere from the soil, it might have ended up in the south Atlantic (or some other ocean), brushed against the water's sticky surface by a random downdraft, popped by its own molecular agitation across the interface, and dissolved.

In the ocean more stories unfold, the details of which must remain untold for now. But they involve an alga, a tiny swimming crustacean called a copepod, the fall of its fecal pellet, a bacterium a mile under the water, the return to dissolved bicarbonate, then a sinuous, lethargic ascent for several hundred years, an equatorial liberation as airborne CO_2, another tour above the seas and continents, and then entry into a flower in a Zen garden in Kyoto, which is picked and arranged for a tea ceremony by a practitioner of ikebana and afterwards dutifully laid atop the temple's compost pile.

Clearly, the byways an individual atom follows are convoluted and unpredictable; the atom never traces the clocklike rounds of blood, the moon, or ferris wheels. But the carbon atom does return over and over to generic forms and places: CO_2 in the atmosphere, bicarbonate in the ocean, and (across geological time scales) calcium carbonate rock. One vast generic form that it assumes encompasses hundreds of thousands of specific and specialized molecules; sub-categories of this form include proteins, lipids, carbohydrates, and nucleic acids. These hydrocarbons and all other molecules within organisms are collectively designated as biological mass, or biomass.

Atmosphere, ocean, soil, terrestrial biomass, marine biomass, carbonate rocks—these places are called reservoirs, or pools. They store particular forms of carbon (often several forms per pool). Just like water reservoirs, the pools of carbon are drained by outgoing flows and are replenished by incoming flows. Thus the storages are not permanent but dynamic. These flows, also called fluxes, connect the reservoirs to one another, forming an overall network of carbon sites and streams. This network offers one way to picture Gaia's body, a picture perhaps as crucial to geophysiology as the diagram in physiology that shows our body's organs connected by tubes of blood, and networks of nerves.

What establishes the dynamics of such a system? In other words, what governs the flows? To move toward an answer, let's first consider

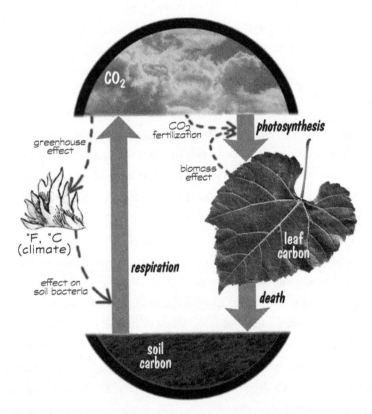

Mass fluxes and causal arrows. Carbon flows in a cycle from the atmosphere into biomass and then back. The fluxes of carbon, shown as solid lines, are regulated by causal arrows, shown as dashed lines.

just two reservoirs of the vast carbon web and the single flow that links them. The picture I have in mind is related to the downward half of the Mauna Loa oscillation. One reservoir is the atmosphere with its store of CO_2. The other is the living biomass. They are connected by a thick arrow going from the atmosphere to the biomass: the flux of photosynthesis.

Photosynthesis clearly depends on the amount of plants, or—perhaps better—the amount of active leaf biomass. The more biomass, the greater the photosynthesis (when all else is equal). Thus the leaf biomass is a causal factor that affects the magnitude of the flux. I find it useful to diagram the activity of a causal factor as an arrow. This arrow—it is distinguished from a flux arrow by being shown as a dashed path in the diagram—links the causal factor to the thing or things it affects. A causal factor may affect a mass flux. It may even affect another causal factor. In this example, a causal arrow wends from the biomass to the mass flux of photosynthesis.

Carbon dioxide also affects photosynthesis. Current journals of agriculture and plant physiology are bulging with results from tests of vegetables, grains, herbs, weeds, and even trees. Their growth gets monitored within a variety of controlled environments, ranging in size from artificially lit chambers to an acre of trellises of gas supply pipes in open fields. The aim: to quantify the effects of CO_2 on plant growth.

For several years I have been collaborating with Bruce Bugbee, a plant physiologist at Utah State University. Bugbee grows wheat at different levels of CO_2 in chambers he designed. An inveterate tinkerer with electronics, Bruce has his house fully rigged with all kinds of sensors. (During New Year's Eve parties, the crowds gather around the computer screen to see whether the carbon dioxide levels in the living room will reach record highs.) In the lab, he finds that the biomass and seed yield of wheat increases by about 25 percent when the wheat is grown under doubled CO_2 concentrations. Such response differs across species (hence the flood of papers to journals). But without getting too bogged down in details, we in the business can confidently state that all else being equal, elevated carbon dioxide boosts plant growth. Thus a second causal arrow should be drawn from the CO_2 reservoir to the flux of photosynthesis.

I am trying to lay some groundwork with explanations of pools, fluxes, and causal arrows because they are crucial for thinking about a global metabolism. At the 1994 Gaia conference in Oxford, convened by Jim Lovelock, one scientist made a remark that lit up my neurons. Mae-Wan Ho, a biophysicist concerned with the inner control dynamics of living things, found herself in the midst of technical talks primarily about global geochemical pathways. "You know," she told me, "I feel perfectly at home, because diagram after diagram—they all look basically the same as those my colleagues and I draw for organisms." The physiologists and geophysiologists are using the same conceptual instruments to navigate similar seas.

The drop of 7 ppm of carbon dioxide in the Mauna Loa graph during late spring and summer is not enough to slow the rate of photosynthesis noticeably. But during chamber experiments at NASA's Kennedy Space Center, I have seen drops that can. Crop physiologists there often want to gather data for a photosynthesis–response curve to CO_2—in other words, to assess the strength of the casual arrow. They switch on the lights inside a twenty-foot-tall, sealed cylindrical growth chamber filled with trays of wheat, soybeans, or potatoes, and then they watch what happens. At first, primed to an initial, high amount of several thousand ppm, CO_2's drop is swift. The descent gets slower and slower, as the photosynthesis becomes less vigorous, and eventually bottoms out at the compensation point, about fifty ppm of CO_2—the point at which the crop's ability to photosynthesize just equals the respiration it must perform to stay alive. This all happens in a day.

Such a crash would take longer for the earth: ten years for the atmosphere alone, not even a thousand with the ocean included. A simple system with two reservoirs and a single flux (and an inherent crash) may be good for experiments but not for the world. The system lacks a primary cycle in the flow of carbon itself. What is missing is respiration. For that, one would add to the diagram a soil reservoir, with hordes of hungry soil bacteria. Supplying this new reservoir is a flux of carbon (in the form of biomass) from the plants to the soil, which represents the flow of all detritus: the autumnal fall of leaves, the toppling of dead trees cracked by the wind, the sloughing of root border cells. Finally, a return flux completes the cycle of carbon. This one loops from the soil back to the atmosphere and arises from the exhalation of the soil microbes.

One more item would round out the set of generic components in this specific picture. So far the causal arrows have originated in the reservoirs: one from the air's CO_2, the other from biomass. Both affect photosynthesis. What about a cause that is not a carbon reservoir? Temperature fills the bill. In the diagram, temperature is represented as a cause outside the reservoirs and is connected to them by two additional causal arrows. One, the greenhouse effect, begins with CO_2 and ends with temperature. The second begins with temperature and could possibly lead to the flux of photosynthesis, but that relation proves notoriously complicated and so will be left blank for now. Instead, a causal arrow is drawn from temperature to the flux of respiration from soil, because it is a well-grounded fact that microbes devour the available detrital carbon faster when the soil is warm.

Now the portrait is complete, with examples of all the essential tools we need to think about the complexity of the total gaian system. This model carbon cycle—its fluxes cycling and regulated by causal loops—can be set to maintain a steady state in which carbon dioxide, plant biomass, and soil carbon are three interconnected pools whose incoming flows balance their outgoing ones. Questions can be asked about the system, such as "What happens when the steady state gets perturbed?"

Of vital importance in answering questions about the dynamics of these systems is the concept of feedback. If a perturbation that increases the amount in reservoir M, for example, loops fully around the system of fluxes, pools, and causal factors and ultimately comes back to increase M still more, then the feedback is called positive (it amplifies a perturbing increase or decrease). If, after the loop, M is less than what it was following the perturbing increase, then the feedback is called negative (it dampens a perturbing increase or decrease).

The little system I've cooked up exhibits both positive and negative feedback and can be used to think about the world today. We certainly are perturbing the atmosphere with our industrial emissions of greenhouse gas. Hence the impassioned quest to understand plant responses: More vigorous plants could dampen the rate of increase from these injections by sequestering some carbon that would otherwise accumulate in the atmosphere. The plants cannot shift the direction of the perturbation from up to down, but they can blunt its magnitude a bit. Also, plants would probably shunt some of the extra biomass to the soil reservoir via detritus, which could further tighten the brake on CO_2's upswing.

Ah, but what about temperature? Warmer bacteria would be livelier bacteria, which would respire more copious quantities of CO_2 from soil to atmosphere. This cycle of causes—from atmosphere to temperature to soil bacteria and their exhalations back to the atmosphere—constitutes a positive feedback loop and thus is potentially worrisome. It would shift some of the carbon held in humus into the air, thus amplifying the fossil fuel injections.

Can the global data shed any light on the integrated situation? In recent years, growing numbers of scientists have begun accepting the real possibility of invigorated photosynthesis on land. For many years, attempts to balance the carbon budget for the atmosphere with firm numbers have been plagued with uncertainties. The amount of CO_2 emitted by the burning of fossil fuels—that's pretty well tagged. The annual atmospheric increase—known to the tenth of a ppm. The increase chugs along at about half the emissions, so half is sinking into the other reservoirs—but where? The tally for the terrestrial biota is understandably one of the fuzziest numbers, but it has seemed to indicate an approximate balance between sources to and sinks from the atmosphere, between lamentable tropical deforestation and the marvelous trend in the temperate latitudes where farms are returning to forests.

What about the ocean? It is definitely absorbing carbon dioxide, but how much? Oceanographers, evaluating the rates of gas exchange, originally thought that the flux driven into the oceans by the rising pressure of CO_2 in the air (the reverse of its escape from root beer) was the remaining balancing factor. But in recent years, they have grown less confident about that.

The atmospheric data often give clues beyond our wildest dreams. Remember that Mauna Loa record of 7 ppm down in spring and summer, then 7 ppm up in fall and winter? Well, it wasn't always 7. Fortunately, we have several decades of records now. The amplitude of the cycle has swelled by about twenty percent since the early sixties. In those initial years of monitoring, the cycle seesawed up and down by a little less than 6 ppm. Right before our eyes, the breaths of the biosphere have been deepening.

Investigators of the carbon cycle have called this change "striking" and "unprecedented." Have the annual dips been dipping ever lower below the trend line because of increased photosynthesis? Have the annual peaks been peaking ever higher because of increased soil respiration? Or is it some combination of factors?

In a recent analysis, Dave Keeling has struck gold again. He and his colleagues have uncovered a correlation strong enough to suggest causation between the increasing amplitude of the Mauna Loa cycle and an upward trend in global air temperatures from 30° to 80° north latitude. What makes them suspect causation is the similarity between the patterns of the two trends, which extends to many details of the wavy, quasi-decadal lulls and spurts. It looks like temperature, as a causal factor in photosynthesis (a topic I deflected several pages ago), is indeed pumping up the biota. If so, a negative feedback loop exists overall between carbon dioxide and the terrestrial ecosystems via the intervening path of temperature.

The experiments with CO_2 fertilization, the missing sink in the carbon budget, the enlarging seasonal cycle, and some other lines of reasoning involving the oxygen cycle and the slight latitudinal gradient of several ppm of CO_2 between the Northern and Southern Hemispheres—these lines of reasoning suggest a plot that is beginning to thicken around an invigorated biota. But we are far from ready to call the mystery solved.

The response to temperature was totally unexpected. Best estimates peg the direct effect of CO_2 fertilization on the Mauna Loa cycle at causing no more than a third of the overall twenty percent increase. So how is temperature acting? No one yet knows. Higher temperature usually promotes increased growth in chamber experiments, but not nearly at the magnitude implied by the correlated trends of the Mauna Loa cycle and Northern Hemisphere temperature.

Keeling at the same time discovered another trend in the Mauna Loa data. As the amplitude has grown by twenty percent, the point at which the descending portion of the cycle crosses the average value has shifted earlier in the year by about a week, while the ascending node has remained about the same. Could spring be arriving earlier? The extra week would boost growth and increase the total photosynthesis for the season. But could that cause most of the twenty percent signal? Furthermore, at Barrow, Alaska, over the same interval, the biosphere's breath has increased by a whopping forty percent, a biological shift even more daunting to explain. Perhaps temperature has driven up rainfall and given growth a boost, but those data are difficult to average on a hemisphere-wide basis. What is clear is that the biota are strongly responding. We inhabit a vast, mysterious, sensitive system. We know that life is responding to changes in carbon dioxide and climate; in turn, life is affecting carbon dioxide and thus climate. We stand witness to the cycle of causation. Now the precise details of the dynamics must be unraveled.

The Prime Directive for Gaian Inquiry

I have so far tried to demonstrate some of the ways of looking at Gaia's body and to suggest a couple of key tools to guide investigations. Tools can be considered either physical or conceptual. The physical tools for making measurements include infrared analyzers for greenhouse gases, mass spectrometers for isotope ratios, computers, satellites. Tools such as mathematics serve on a more conceptual level in the construction of knowledge: to build data into systems, to quantify correlations in trends, to build formal models of reservoirs, flows, and causes. In formal models, which I make all the time, one writes equations for the parts of the system. For example, in the system discussed in this chapter, one would need equations for carbon dioxide, leaf biomass, soil carbon, photosynthesis, detritus fall, soil respiration, and the greenhouse effect.

But how do the guiding ideas originate? What generates ideas for the diagrams whose details will be informed by the measurements and formalized by the mathematics? Answering that would require disentangling the constituents of inquiry itself, and I'm certainly not going to attempt to untie that Gordian knot in this paragraph. Nevertheless, I will suggest that just as hammers and saws enable us to build any number of physical things, there are probably conceptual tools working deep in the mind to construct various models as we seek to build understanding. Morphing the metaphor only a bit, these conceptual tools become vehicles—like sailboats, bicycles, rockets—that give us the flexibility to visit any number of different places. In gaian inquiry, the conceptual vehicles are the directives for thinking about the whole Earth.

One directive would beseech us, "Attend to the cycles of matter." Pedaling with this vehicle yields several advantages: The tour encircles the globe; the perspective provides an overview from which to ask more detailed questions about any subsystem; the animate and the inanimate get bound together into a system. Carbon as lead actor and oxygen as walk-on have dominated the scene so far, but also crucial are nitrogen, phosphorus, sulfur, calcium, and other elements, including life's essential trace elements such as iron and magnesium. Their global circuits are called the biogeochemical cycles. These combine biology, chemistry, and geology. Studying them takes us directly into the guts of Gaia's metabolism.

A second directive certainly follows: "Attend to the cycles of causes (of the cycles of matter)." With the discovery of the double circulation of the blood, physiology was off to a great start. But eventually the practitioners of biology and medicine wanted to know more: What controls the circulation rate? What makes blood pressure high or low? How did the blood system evolve? Similarly, the cycle at Mauna Loa opened our eyes to the dynamics of the biosphere. Questions about causes soon followed: What controls the rates of photosynthesis and respiration at large scales? Can imbalances occur? How is the system affected by population shifts in species and by evolution? As we saw in the example described earlier, the pools within any biogeochemical cycle are often causal factors themselves. Other causal factors stand outside the reservoirs, as temperature did in our example. The arrows that lead from causal factors to effects are simple paths with beginnings and ends. Yet together, interlaced with the reservoirs and fluxes, the causal arrows shimmer in a network of simultaneous feedback loops.

Embarking on this voyage can sometimes make one feel one is far upstream in the heart of darkness. In journals such as *Microbial Ecology*, *Agroforestry Systems*, *Global Biogeochemical Cycles*, and *Plant Systematics and Evolution*, to name just a few of hundreds, so much seems potentially relevant to geophysiology that one is apt, like Kurtz, to start murmuring "the horror, the horror." Or "the complexity, the complexity." The dynamics in just the system of photosynthesis discussed in this chapter can get mind-boggling, as I found out after doing half a dozen pages of math to explore the impacts of various assumptions—all reasonable—about the mere handful of causal factors. And this system was truly simple. It ignored distinctions between grasses and trees and among the vertical layers of soil; it pretended the ocean did not exist, not to mention the probable need to consider Earth's oceanic regions of varying climate and biology. When one gets to the real Earth, what to do?

Scientists naturally want to keep measuring, filling in ever more details of their construct, and refusing to rest until every board is cut to a fraction of an inch tolerance and properly nailed. That's fine, and experts are often needed at that stage. But everyone can admire basic house design, just as everyone can revere the whole organism before dissection begins and can appreciate a view of the canyon before hiking down to the individual boulders. That means thinking about big principles. What are the big principles of geophysiology that unify the complexity of causal arrows, reservoirs, and fluxes?

One possibility may be borrowed from ecology, the concept of keystone species. The term derives from the keystone of an arch, the wedge-shaped stone at the summit that establishes pressure downward in both directions and so locks the arch in position. For an example in ecology, muse on what beavers do to their surroundings. Their dams create ponds and swamps, raise the water level, and slow down drainage throughout the area. By increasing the availability of water for other living things, the beavers affect the entire ecosystem. Moreover, keystone species are not always the big critters. Nitrogen-fixing bacteria can be crucial in low-nutrient soils by injecting a flux of ammonium into the web of life.

For the overarching system of Gaia, then, there may be another directive: "Find the keystone parts in the arches of matter and causes." Certain reservoirs could be vital because of their immense size, acting across the entire system to buffer perturbations. Particularly vital fluxes are also a definite possibility.

Like the nitrogen-fixing bacteria, the key parts in the gaian system may not always be those that first strike the eye. Recall that of the substantial amount of carbon dioxide in the soil's air, only about one molecule in a hundred ends up as bicarbonate in rivers flowing to the sea, carrying along calcium ions liberated from continental rock. This minuscule flux proves crucial to determining the atmospheric level of carbon dioxide on geological time scales.

Though it turns out that land plants drop from the picture as a carbon-storage pool on these time scales, they will have much to do with the changes in carbon dioxide during the next century (they are significant already). Thus various reservoirs and fluxes change in importance as we adjust our temporal focus. This fact adds another to the growing list of directives for gaian inquiry: "Attend to time scales." The breaths of living things create some of the most rapid cycling times within the biogeochemical cycles. But across the vistas of eons, the scale of inquiry would shift to biological evolution itself. Gaian inquiry should thus focus on how the evolution of certain types of creatures has altered the biogeochemical cycles.

Gaia is not a system of complex but ultimately dead feedback loops, like the capacitors and resisters in an intricate electronic circuit. Gaia contains living beings. From the interaction of life with the inanimate forces of nature emerges a sort of global symbiosis. The interdisciplinary scientist and philosopher Gregory Bateson wrote about ecology as a mental phenomenon. He saw a similarity between the loops of causes in ecosystems and those in nervous systems. Parts of the gaian system are organisms with sensitivity and responsiveness. We do design responses into the parts of our electronic wizardry, but the difference is that for Gaia, the responders have evolved on their own within the matrix of the whole. The plants do not merely accept the onrush of CO_2 when that gas reaches higher concentrations in the atmosphere. Their various responses have evolved. Thus even the causal factor that runs from the reservoir of CO_2 to the flux of photosynthesis must be considered, ultimately, from the viewpoint of evolution.

Attend to the cycles of matter. Attend to the cycles of causes of the cycles of matter. Attend to those reservoirs, fluxes, and causal factors that occupy especially key positions within the entire network. Attend to the unique combination of life and nonlife. Attend to time scales, noting how different parts emerge as crucial depending on the temporal frame of observation, and especially, over the longest frames, attend to the interaction between the cycles and the evolution of life. Putting together all these directives, I contend, provides a set of conceptual tools—a single mental vehicle—for exploring Gaia.

In its early days, James Lovelock called his idea the "Gaia hypothesis." The specifics in his writings have developed over time, but I think it is fair to summarize the hypothesis (without any mysticism) as postulating a self-regulating superorganism. Various pieces of evidence for the hypothesis have accumulated over the years, as substances that may enhance this self-regulation have been discovered and as models of feedback loops have demonstrated the concepts. The ups and downs of the Mauna Loa cycle, for example, stand before our eyes as evidence. Lovelock now prefers to talk about "Gaia theory," which is a level more logically broad and encompassing than any single hypothesis. I agree with the enlargement of logical levels, but I prefer to talk about "hypothesis generators."

Just as trees are fewer in number than the leaves they bear, so hypothesis generators must be fewer than hypotheses themselves. Where do hypotheses come from? Do the minds of scientists (and everyone else) contain hidden hunches, based on deeply held mental images about how some entity works, which they then use to guide the particular questions that come into consciousness? In the case of Gaia, can an underlying generative idea be at the heart of the list of directives? If I had to declare one central driving force behind all the directives, what would I say? It would have to be something that would give most insight into the workings of the planet, a core method for investigating, certainly something related to life. It would have to be a focal point around which the other directives revolve. This prime directive might be: "Think about how the planet would be different without life or particular forms of life."

With and *without*: The words form a binary system, the essence of science. Scientists (and not only scientists) simplify complexity for ease and precision in exploration. For example, the classic procedure of the scientific method compares an "experiment" to a "control." The primary question of gaian inquiry—What would the earth system be like without life?—forces the investigator to consider a binary: Earth with and without life. Performing the comparison will tell us something about the workings of Earth by creating an experiment and control—not in reality, of course, but in concept. This act is right at the heart of the scientific method.

The binary comparison of Earth with and without life (or some forms of life) has been central to the work of many of the scientists seeking inspiration from the gaian paradigm, including the founder himself. The original hypothesis grew out of Lovelock's work with NASA in the 1960s, from his cogitations about a pair of planets. Lovelock was charged with designing experiments to detect life on Mars. How does the atmosphere of Mars compare to that of Earth? Mars has a tiny amount of oxygen and no methane. Earth has thirty thousand times more oxygen than Mars and a smidgen of methane. The methane proves crucial because even that smidgen could not exist in the presence of so much oxygen were it not for life, which continuously produces methane. Furthermore, the oxygen on Mars can be accounted for abiotically. Such considerations played a pivotal role in Lovelock's early predictions of a currently lifeless Mars, predictions based on what life must inevitably do to a planet's atmosphere.

Many papers about Gaia have invoked a basic binary comparison to say something about the strength of life in the Earth system. What would clouds be like without the cloud-seeding gas, dimethyl sulfide, that is emitted by plankton? What would carbon dioxide levels be without the enhancement of chemical weathering by ancient land life as bacterial crusts? Or without additional enhancement later by rooted, soil-securing trees? How would Earth's climate over four billion years have been different without life? How would the nutrient cycles of sulfur, phosphorus, and nitrogen have been different? Would they have rolled at all? I say to you god or gods, "Yahweh, Lord, Allah, I do not yet know." I say to you goddess of the Earth, "Gaia, I want to find out. And, in the meantime, help me formulate my best opinions." Some of these topics I have worked on, and I feel that these questions are among the most fascinating in science today.

Without life, the annual graph of carbon dioxide at Mauna Loa would be as flat as the EEG of a dead body. Or, if an annual seesaw did exist in the absence of life, it would have to be driven by the ocean's thermal cycle. The oceans may outgas like root beer when they are warmest, in summer, but that flux would run counter to what is observed. Life produces a cycle in this case quite opposite from that of thermodynamics. As far as we know, no other process on Earth could generate cycles of carbon dioxide that decrease in spring and summer and increase in fall and winter. The Mauna Loa oscillations are thus a sign of life, borne as a signal in the atmosphere, carried around the world by the winds—a glimpse at photosynthesis and respiration on a global scale, two great processes of Gaia's body.

2
A Global
Holarchy

Recently I visited the place of my childhood, North Tonawanda, an industrial town ten miles upstream from Niagara Falls, between two of the largest freshwater lakes in the world. On a mid-May morning amidst glorious sunshine, I ambled around the perimeter of my elementary school, a 1950s tan-brick, stand-alone classic. The old gaming area for marbles—trammeled dirt and scattered pits that challenged the most cunning strategies of a nine-year-old—was now smooth with green lawn. I guess kids don't let passions fly with cat's eyes anymore. Before heading back, I detoured to pay my respects to what for me had once been Big Nature, the nearby Pine Woods Park.

Leaving suburban civilization and crossing the street brought me first to a transition zone of several ball courts, kept mowed by the town's maintenance crews. The manicured fields terminated abruptly at a towering wall of trees, where began the central, grandest, and formerly wild portion of the park. For the last couple of decades, such visits had always evoked a sweet but bitter nostalgia. Sweet because of the communion with nature I long ago learned there. Bitter because wildness had since been subdued and defiled. About twenty years ago, a violent crime (rare in that town) precipitated a change in management policy. Annual clearing operations removed the wild undergrowth. In contrast to its natural, vast mysteries, with tunneling paths for childhood romps, the park became transparent to the traffic and homes on all sides.

But on that day in May—wow. Because funds had shriveled, because memory of the crime had faded, or because an environmental group had interceded, the park's jungle was making a luxuriant comeback. Suddenly I could get lost again. And I did.

My joy during the jaunt through the reborn woods could not be contained. "Pine Woods Park is back! Pine Woods Park is back!" I shouted. I was lumping all the creatures of the park into a unity, a glorious being of its own. A green phoenix had risen. Not just all the myriad organisms had returned; *it* had returned. Of course I noticed the individuals included within the resurrection, the patches of brilliant white trilliums, the exuberant vines, the scampering squirrels, the hopeful, slender young birches, oaks, and, yes, pines. But mostly I was moved to joy by the single entangled whole.

In less ecstatic moments, as well, I would call the park an entity. This unit of land management within the town clearly ranks as a thing because its legal borders are visibly defined by the surrounding streets. Three hundred years ago, when the entire region was more or less forested, that particular plot would not so obviously have deserved a name. Is the possession of borders crucial to a thing's being a thing?

Here I'd like to introduce a word used in my book *Metapatterns*. I borrowed a term—*holarchy*—from systems theorist Arthur Koestler. A holarchy consists of any whole and its parts. These parts, in turn, are wholes themselves with their own parts, and so on. Thus a holarchy can be considered as the nested system of wholes and parts over numerous levels. Pine Woods Park is a holarchy of ball courts and wilderness, with the wilderness in turn containing its living multitudes.

Moving outward from Pine Woods Park in lateral fashion, one can pass through a nested series of ecological or political units, eventually reaching Gaia. But the quickest route to the global scale is upward. Including the atmosphere in the next layer of the holarchy of which the park is a part immediately connects it to all other life, as well as to all soils and the ocean. The story of such connection can be told, as we saw in Chapter 1, by following the twists and turns of carbon. It can also be told by tracing the pathways of the air's most abundant gas.

MICROBES AND GLOBAL NITROGEN

To understand nitrogen in the atmosphere, one might well begin in the ocean, even though that seems an unlikely place to start. Moreover, the story will include carbon and sulfur and microbes in muck—all necessary to make visible the coupling between life and its chemical environment, including the atmosphere.

Let's begin in the ocean by joining a fishing ship out from the Bay of Concepción, along the coast of central Chile at 36° south latitude. The prolific life there yields terrific harvests of shrimp and several types of fish with hooked lower jaws, called hake. The plenitude of fish and shrimp is due to the plenitude of what they eat; the hake themselves eat shrimp as well as sardines and anchovies. These little fish feast on the cornucopia of zooplankton, which feed on the countless tons of green phytoplankton cells. But what accounts for the richness of phytoplankton? As in all marine waters, there is plenty of carbon dioxide in the form of bicarbonate (HCO_3^-). No scarcity there. Ocean life is usually limited by the dissolved nutrients, such as phosphates and nitrates. We're in the right place for a wealth of those, too. The western coast of South America is blessed with some of the most famous areas of coastal upwelling in the world.

Fish generally flourish in coastal regions. Oceanographers speak of "green waters" in contrast to the relative deserts of open ocean, the "blue waters." The enhanced mixing along coasts, where seas meet continents, keeps nutrients stirred up from the bottom. Most remarkable of all are those coasts supplied by deep upwelling, such as the western sides of South America and Africa. There water is not just stirred but vigorously pumped upward from the deep ocean by winds and the geometry of the land. The upwelling waters are laden with nutrients. In only a few tenths of a percent of Earth's ocean area, these coastal upwelling regions produce half of the harvested fish.

Even so, the fishermen do not exactly unroll their nets, put their feet up on the railing, have a smoke, and wait for stocks of fish to jump in. Profitable harvests mean hard work. Off the Bay of Concepción, the ships trawl for the shrimp and hake. There the ocean bottom is not far below. The continental shelf—submerged but geologically a part of the continent—extends from the coastal port of Concepción to about thirty miles out. At that point the sea floor plunges. But for those miles of shelf, the water is shallow, gradually reaching depths of only several hundred feet. A slope of ten feet a mile—that's not much.

As the nets drag, they can accumulate a messy, twine-like substance called by the Spanish word for uncleansed wool, *estopa*. Estopa is actually the microorganism *Thioploca*. The fishermen might just as well translate that as "nightmare" for the havoc it sometimes plays with their nets.

Thioploca is a bacterium about two-thousandths of an inch long. It becomes visible by growing into colonies of enormous filaments, thousands of cells in length. Filaments gather themselves into bundles of up to a hundred with a common border of a thick mucilage sheath. The resulting strands are the "wool." This holarchy runs from the visible to the microscopic: The sheathed strands consist of filaments, and the filaments consist of individual cells. As wholes, the sheathed strands might reach five inches in length and a sixteenth of an inch across, like half-lengths of overcooked spaghetti noodles. They grow almost vertically, and when conditions are right, an inch of living filaments sticks up out of the sheaths embedded in the sediments. Crucially for this organism's

Strands of *Thioploca*. The strands consist of numerous filaments bundled inside mucilage sheaths. Widths are about one-sixteenth of an inch. Photograph by Markus Hüttel.

physiology, the strands can also glide downward into the underlying layer of black sediment.

The mystery embedded in these mats of microbes concerns how they are so productive in waters nearly devoid of oxygen. At a depth of just a hundred feet, the concentration of dissolved oxygen is only a couple percent of what it is at the surface. The reason for this oxygen starvation is a subsurface current that travels from a low-oxygen region of the world ocean. But oxygen is always somewhat depleted below the ocean's surface—a general pattern of ocean chemistry. To understand why requires a review of the carbon cycle in the sea.

The marine carbon cycle is closely analogous to that on land, which was discussed in the first chapter. In place of dark soil beneath plants, though, the ocean has dark water under the well-lit zone of photosynthesis. Within the uppermost, so-called photic zone live diverse species of algae. From the growth of these phytoplankton, from those that feed upon them, from the ones that feed next in line, and so on, detritus rains

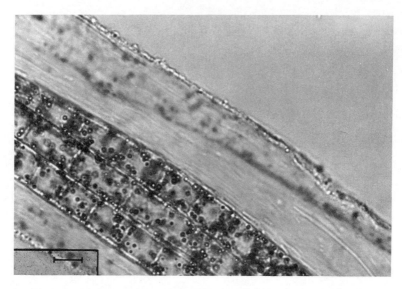

Cells of *Thioploca*. The individual cells within three filaments contain tiny globules of sulfur. A mucilage sheath, about 100 microns thick, surrounds the filaments. Black bar shows 40 microns. Photograph by Markus Hüttel.

downward into the darkness. The detritus consists of dead algae, fecal pellets from all the pellet makers, fish scales, and assorted debris. Bacteria and other creatures find this fall of organic material irresistible. They gorge themselves on the waste and, by their respiration, use up some portion of the oxygen in the water. This dark realm—the ocean's "soil"—extends down hundreds of feet to the base of the water column on the continental shelves and down several miles in the open ocean, from just beneath the photic zone all the way to the bottom benthos.

For a 3000-mile coastal strip where *Thioploca* is found, the subsurface oxygen level is extremely low. Yet an abundant rain of potential food from the surface cornucopia reaches the bottom, where a thick carpet of spaghetti microbes is found. How do they live? These unusual conditions enticed a German team from the Max Planck Institute for Marine Microbiology to work with Chilean scientists on board a Chilean research vessel to sleuth out the mystery in March of 1994 (late in the austral summer). The scientists took samples of the "uncleansed wool" by coring into the sediments. They measured the chemistries of water and sediment and transplanted *Thioploca* mats into aquariums that mimicked natural conditions, to experiment, observe, and expose the mat's metabolic tricks.

One clue was found in the underlying black gunk, for a *Thioploca* community is far more than just *Thioploca*. Some unnamed microbial residents in the underlying muck lead the way in the community's metabolism without molecular oxygen. They use a substitute for oxygen: sulfate. Sulfate (SO_4^{2-}) is one of the most abundant dissolved ions in the ocean, and it diffuses along with the water down into the tight spaces of the sediments. Although oxygen as a dissolved gas in the standard form of two oxygen atoms is not available, oxygen as an element occurs in the sulfate.

The nameless denizens of the black layer are called, as a group, sulfate reducers, which is a fancy way of saying that they can gain energy by splitting sulfate for its oxygen in the presence of organic detritus. (In this process, sulfur gains electrons, which *reduces* its positive charge and explains why the bacteria are called sulfate reducers.) Employing the oxygen they gain in this way, these sulfate reducers feast on the carbon-rich detritus, excreting carbon dioxide as a waste gas. The released sulfur is combined with the hydrogen ion that is present as an acid complement to the sulfate. The resulting hydrogen sulfide is a waste gas to the microbes and is excreted.

One rule that aptly summarizes the science of ecology is that everybody's waste is somebody else's food. That's true here too. Hydrogen sulfide, waste to those sulfate-reducing microbes, contains potential chemical energy for others that have the means to unlock it. Another ion with the potential to unlock the energy in the hydrogen sulfide diffuses with water into the black sediments, but this ion — nitrate (NO_3^-) — is less abundant than sulfate by a factor of a thousand. On the other hand, because the supply of water above the sediments is, in practical terms, infinite, the supply of nitrate is also infinite despite its low concentration. If only some creature could bridge the distance between the hydrogen sulfide within the sediments and the overlying nitrate in sea water. But the copious nitrate available just above the sediments might as well be a universe away to the ordinary bacterium.

Thioploca is no ordinary bacterium. Though one microbe alone is far too small, thousands joined as filaments can bridge the distance. The filaments bundled into sheathed strands create the necessary strength. Within the sheaths that function as commuter tunnels, the interior filaments of living "wool" glide up and down, like subways, in and out of the sediments. Even more, perhaps, they resemble wires stretched between the poles of a battery, from the hydrogen sulfide to the nitrate, except that these wires move between the poles, so each interior cell can gather energetic juice from both sides. The power thereby produced fuels their amazing growth upon the settling rain of detritus. When the scientists doused the aquarium experiments with extra nitrate, they observed *Thioploca* filaments as "white hair, swaying in the anoxic or hypoxic water to provide optimal conditions for nitrate uptake."

Everything is now accounted for. The sulfate-reducing microbes in the black layer take dissolved ions of hydrogen and sulfate, add detrital carbon, and spew out the gases carbon dioxide and hydrogen sulfide. The spaghetti colonies of microbes in the overlying yellow-brown sediments take the hydrogen sulfide and the water's natural nitrate ion, add detrital carbon, and spew out more carbon dioxide (carbon plus oxygen), sulfate (sulfur plus oxygen), and water (hydrogen plus oxygen). All is accounted for, that is, except the nitrogen atoms from the molecules of nitrate. What happens to them?

That is still unknown. Nitrogen might be excreted either in surplus ammonium ions or in molecules of nitrogen gas. The community in which *Thioploca* lives is chock full of unknowns. For example, more sulfate is reduced than can be accounted for in the pathways taken by the resulting hydrogen sulfide. Some of the sulfur in the sulfide is taken up by particles of iron, but how the sediments are stirred with enough vigor to maintain the iron uptake is still a mystery. Perhaps *Thioploca*'s vertical commutes provide the required mixing. The answers to these questions about microbial communities off the coast of South America will affect our understanding outward in the holarchy of life, all the way to Gaia itself.

We can see in the *Thioploca* community that a directive for gaian inquiry that enjoins us to follow the cycles of matter will require knowledge of the simultaneous movements of many different elements. Fluxes within the cycles—and thus the cycles themselves—are linked. The organisms below *Thioploca* use sulfate to oxidize organic matter that carries carbon. *Thioploca's* metabolism is intimately tied to nitrate uptake and the hydrogen sulfide produced by its neighbors. Such linkage extends beyond correlated fluxes. The metabolisms of the benthic microbes depend on the water's level of oxygen, specifically the lack thereof. Here the absence of one element—oxygen—profoundly affects the flows of others.

If *Thioploca* excretes nitrogen in the form of ammonium ions (NH_4^+), these will diffuse in the water and ultimately serve as feedstock for other microbes called nitrifiers. The nitrifiers might live nearby, if they can scavenge enough precious oxygen to perform the nitrification of ammonium back to nitrate. Or they may live at the surface where oxygen is abundant. Exchange of materials among organisms is fundamental to the physiology of Gaia. Though we usually think of connections between organisms as one feeding upon another (hummingbirds on flowers, owls on voles), the links within Gaia depend more on the chemical exchanges among organisms via the surrounding media, such as sea water. Sulfate, hydrogen sulfide, nitrate, carbon dioxide— these are the packages whose transfers create the networks of Gaia's metabolism.

If *Thioploca* excretes nitrogen not as ammonium but as nitrogen gas molecules (N_2), the spaghetti microbe will have joined the ranks of what are called denitrifiers. Whether or not *Thioploca* itself is a denitrifier, it lives in a region teeming with them. Microbial denitrifiers in the sediments and in the water column survive by using nitrate as an oxidizer in places where their potential competitors are inhibited by the low levels of oxygen. We know denitrification happens, and we know it happens in a big way in the eastern Pacific. In that region, measurements of the ocean's chemistry reveal the presence of denitrifiers by telltale depletions in the concentrations of nitrate. Yet denitrification is one of the most enigmatic processes in the ocean. Studies using genetic probes suggest that only about one percent of the marine denitrifiers have been characterized.

In denitrification, nitrogen is transferred from its usable form (biologically speaking) to the relatively inert form of nitrogen gas, a process that moves this tale's conceptual pathway out of the water and at last into the atmosphere. Nitrogen gas excreted by marine denitrifiers escapes the sea and enters the atmosphere. Imagine nitrogen gas bubbling up out of the waters all along the coast of South America and hundreds of miles out to sea. Denitrification occurs terrestrially, too, of course, wherever some microbe regards nitrate as an oxidizer. As in the sea, denitrification on land takes place where the concentration of oxygen gas is vanishingly low. And what is the land's biogeochemical analog of the deep ocean? Yes, the soil.

Wherever air percolates too slowly into particular depths or patches of soil for aerobic life to survive, there anaerobic microbes, if they denitrify, can feast on anything organic that comes their way, such as old, unrecognizable remnants of fallen leaves. The denitrifiers of the soil (all bacteria) include several species of the common genus *Pseudomonas* (which once gave me a nasty ear infection), as well as more exotic, specialized species. Compared to the sea, the vertical distances are puny, but the biological and chemical pattern is much the same.

Given all the fluxes from denitrifiers in seas and soils, one might expect the atmosphere's nitrogen content to increase without bound. As we shall see, however, the volume of nitrogen in the atmosphere is already about as high as it can get, for the simple reason that very little of it exists anywhere else. Even so, denitrification by itself would further deplete the sea and soil of precious nitrate. To understand the dynamics of the nitrogen cycle, an opposing flow must be considered. The vital, balancing flux is that of nitrogen fixation.

To a pine tree or grama grass, nitrogen gas in Earth's air might as well be on Mars. No plant—no plant whatsoever—can break the triple bond between the two nitrogen atoms and convert the freed atoms into a biologically useful form, such as ammonium or nitrate. But special, fix-it bacteria can accomplish the job. By way of unique enzymes, they sever the bond between the two atoms of nitrogen in the gas molecule and shunt each into ammonium ions. These nitrogen fixers may be free-living floaters at the ocean's surface, or they may be attached to soil particles. Most famously, some form symbiotic unions with the roots of particular plants, an activity crucial to agricultural fertility. Such plants include peas, clover, beans, and alfalfa. Ginkgo trees and alder can also harbor these talented bacteria on their roots. Wild plants that can host the nitrogen-fixing nodules are often the first to colonize generally in-hospitable ground, such as sandy banks washed clean of humus and soil after a river flood.

The ammonium made by nitrogen fixers satisfies the nutrient needs

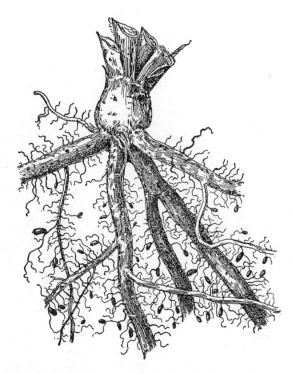

Nitrogen-fixing nodules. Swellings on the roots of red clover house multitudes of nitrogen-fixing bacteria.

of land plants and of algae in lakes and oceans during protein synthesis. Other bacteria, the nitrifiers, fairly quickly convert ammonium into nitrate, which also suffices for the protein-building cellular machinery of plants and algae. The marine cyanobacterium *Trichodesmium* fixes nitrogen and thereby directly promotes its own capability for photosynthesis. Overall, the biological flux of fixation that transforms nitrogen from atmospheric gas into its usable forms in soil or water is the main flux from the atmosphere that counterbalances the flux of denitrification. With regard to the atmospheric pool of nitrogen gas, fixation is a sink and denitrification is a source.

To this material cycle of nitrogen between gaseous and fixed forms, we can apply to one side at a time the prime directive of gaian inquiry ("How would the planet be different without life?"). Without the fixers, the denitrifiers would boost the air's nitrogen. But 99.96 percent of total nitrogen that is not in rocks already exists as N_2 gas, and 99.4

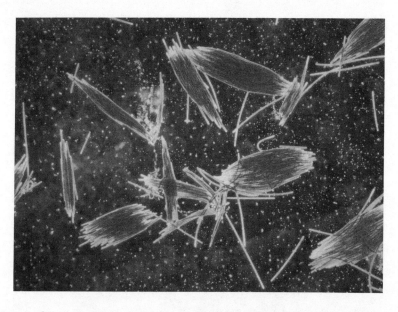

Nitrogen-fixing cyanobacterium. Filaments of *Trichodesmium thiebautii* are shown here in groupings called tufts. Each individual filament (several of which are by themselves) is a chain of about 100 cells. As a genus, *Trichodesmium* is one of the most significant nitrogen fixers in the ocean. Photograph by Hans Paerl, University of North Carolina at Chapel Hill.

percent of that wafts free in the air (the rest is dissolved in the ocean). The remaining 0.04 percent of total nitrogen, which includes every form other than N_2, is distributed among all the ocean's waters, all the soils, and all the terrestrial and marine biota. These numbers are astounding. Apparently the denitrifiers have reached their limit and are maintaining it.

Now consider the other side of the cycle. In a world with no denitrifiers, only fixers, the nitrogen would gradually be taken out of the air and put into its fixed forms, which are highly soluble in water. Fixed nitrogen in terrestrial ecosystems would slowly but surely leak into ground waters and then rivers, and ultimately reach the sea. In fact, the solubility of nitrate in water is so high that nothing precludes the possibility of all nitrogen ending up as nitrate in the ocean. A biota with today's amount of nitrogen-fixing bacteria, but without denitrifying bacteria, could deplete the air of nitrogen in less than thirty million years.

But would it? Geochemists tell me that as the ocean's fluid slowly percolated through the hot, mid-ocean ridges, over about ten-million-year periods, the nitrate would probably be split. Its oxygen would be snapped up by elements in the rock, such as iron. Its nitrogen could then be released as nitrogen gas. Today this process is minor, because the highly efficient biological denitrification allows very little nitrate to remain in the ocean. But the denitrifiers may have a back-up system in Earth's deep heat underneath Gaia.

The thermal back-up, however, would not be as efficient as the biological denitrifiers. Slightly less nitrogen in the atmosphere might be no big deal. But what was slightly less to the atmosphere would be an enormous—at least a hundred-fold—boost to nitrate in the ocean. Consider the impact of this shift from the perspective of evolution. Billions of years ago, nitrogen fixers probably evolved in response to an ongoing crisis in the amount of nitrate in the ocean—a crisis in fact caused by the denitrifiers, which were even more ancient. Imagine this scenario: Without the denitrifiers, nitrogen fixers would not have evolved, because in the presence of a hundred times more nitrate, their alchemy would have been useless. With apologies to the American Revolution,

I therefore propose a slogan to help us keep in mind both the evolutionary sequence and the balance of the entire nitrogen cycle: "No fixation without denitrification."

Who cares? The trilliums, vines, pine trees, and oaks and thus the squirrels of Pine Woods Park, for starters. Nitrogen-fixing bacteria on land came from marine nitrogen fixers. If marine nitrogen fixation had never evolved because there were no denitrifiers—and thus more abundant natural nitrate—in the ocean, that might have been fine for all the ordinary phytoplankton, but it would have limited life on land to a bleak, malnourished existence. Land life would have depended on the trickle of fixed nitrogen derived from lightning. But in the ocean this trickle would maintain an equilibrium level with a (relatively) enormous quantity in storage, supplying plankton, via upwelling, with more than they could use. On land the trickle would constantly wash away down rivers. The lush savannas and rain forests that grace the planet would never have evolved.

In denitrification and nitrogen fixation, we see the visceral biological forces at work and the intimate union of life with the fluxes and pools of the global metabolism. In the end I think it is denitrifiers that are to be celebrated as key gaian agents, if only because they are less well known than their more famous partners in the cycle. However the story of evolution unfolded, it is awesome to consider—with a deep breath—that seventy-eight percent of the volume of air we inhale has been excreted as a waste gas by bacteria that thrive only where oxygen is absent.

Outward and Inward Influences

What do all these details about the nitrogen cycle have to do with Gaia? Perhaps the grand understanding must be harvested one fruit at a time. Will the details eventually fill the bowl, adding up to everything we need to know about how the planet works? In the meantime, we can at least nibble on the tidbits of insight into the Herculean labors performed by mighty microbes, by plants breathing the air, and by fungi cleansing the soils.

On the other hand, we may be ready for more than nibbles before the whole bowl is full. Can satisfying generalizations be had? Our investigations have yielded fruits as we explored the carbon cycle, the nitrogen cycle, and even that of oxygen. One of the directives for gaian inquiry has been followed: Attend to the cycles of matter. The prime directive has dutifully been employed. What more can be learned?

To discover something general about the structure of Gaia, it will be helpful to spend some time thinking about Gaia as a holarchy. Looking at any holarchy raises questions about the insides of things and their outside contexts. For example, a fox is a holarchy that surrounds metabolic and chemical nestings of organs that have tissues that have cells that have organelles that have molecules that have atoms that have subatomic particles. One can move outward as well. That same fox is within a holarchy of the fox family in the local fox population in the ecological community. There are many ways of looking at the nestings, and this topic alone could fill a book. For now, let us accept that Gaia clearly is a holarchy.

What are the other levels in the gaian system? Examining the parts of Gaia is one obvious avenue of inquiry, for in that direction lies all the intriguing complexity of rain forests and foxes and bacteria. But there is another direction: With what other things does Gaia itself form a larger system? This question leads right away into a very real problem that emerges whenever one is tempted to draw comparisons between the workings of Gaia and the workings of organisms — a problem that can best be seen by expanding outward, first from organisms and then from Gaia. Let's go far out.

What is the system within which our bodies are parts? It could be the family, the community, the nation, the bioregion. Many answers are possible. One incontrovertible answer is that you and I are both parts of a gene pool. The individuals of our species evolved into the shared humanity we see in the world today by having been biological parts within a genetic system (whether one or several gene pools does not matter here). Call the organisms in a shared gene pool a system of evolution.

The key structural concepts for a system of evolution are heritage, variation, and natural selection. The heritage comes from having a population of individuals that are very much alike and ultimately connected to a shared ancestral stock. Variation comes into play because organisms of a species differ across a wide spectrum of details. Natural selection implies that the viability, reproduction rate, survival rate, or other properties of the organisms are filtered selectively, over time, on the basis of this variation. This threefold process of heritage, variation, and natural selection, over enough generations in interacting systems of evolution, turned dinosaur arms into bird wings and bulked up the human brain.

One can comfortably generalize the overall idea of a system of evolution. Any population of similar things can be said to evolve if successive generations vary across a spectrum of details, and if this variation triggers a process of differential selection that creates change in the average member of the population over time.

For example, a number of years ago I bought a mountain bike. Within a year, the manufacturers had changed the gear shifts and the hand grips. Products in free markets are systems of evolution, because technological improvements and consumer preferences can make products evolve. As another example, consider those Little League teams batting it out in the ball courts of Pine Woods Park. The Mounties, Padres, Indians, and Angels, as individuals in the population of teams, should be considered a system of evolution, because strategies such as batting lineups and practice schedules shift over time, based on the coaches' observation of kids' performances.

Compared to entities that are parts of varied and selected populations, such as a baseball team, a bicycle, or an organism, Gaia is in a league by itself. Even by the broadest possible definition of a system of evolution, Gaia is not a part of such a system. There is no population of Gaias. There is only one Gaia, self-contained by necessity, alone in the universe so far as we know, or, if not alone, not in communication with the others, unless it turns out that fragments of alien genomes are raining down in meteoroids from space.

This profound difference between our bodies and Gaia has been most forcibly voiced by evolutionary biologists. Pondering this difference yields many insights into the structure of Gaia, both in space and in time. For example, though Gaia is not itself part of a system of evolution, it does contain numerous systems of evolution, as least as many as the tens of millions of species.

The larger holarchy that contains Gaia must, as we will discuss later, include the other parts of planet Earth: the rocky crust, the upper and lower mantle, the core. The larger holarchy must include space as well—its meteoroids and, most prominently, the luminous sun. To the extent that Jupiter affects the flux of comets and our moon governs the tides, the larger holarchy must include these other astronomical bodies too. But again, none of these parts of the larger level seem anything like Gaia; they do not contain living beings in evolutionary systems as their parts, as Gaia does.

So what, then, is Gaia? This challenge might be approached by delving slightly deeper into the theory of holarchies.

One person who shed some light on holarchies was Jacob Bronowski. A physicist, mathematician, and biologist, he is remembered most for his passion in bridging disciplines and relating art, philosophy, and human ethics to the practice and ideas of science. Bronowski wrote such classics as *Science and Human Values* and *The Ascent of Man* (he presented the television series). In 1970 he published an article titled "New concepts in the evolution of complexity: Stratified stability and unbounded plans." Here Bronowski distilled from the complexities of all reality a general pattern, which I interpret as a major insight into the relationship between parts and wholes. Bronowski noted that molecules, being systems of atoms, could have come into existence only after the atoms themselves. Similarly, cells could have come into existence only after molecules. Cells are not built directly from raw atoms; the evolution of cells required an environment that included the intermediate stratum of molecules (or, better, the several intermediate levels of simple molecules, complex molecules, and highly reactive molecular complexes).

Multicellular organisms—animals, fungi, plants—though they contain molecules (for example, the bicarbonate ions in our blood), were assembled not from molecules but from cells. The cellular layer of complexity had to precede that of the multicellular organism. And not just any type of cell would do, but only the so-called eukaryotic cell, which sports a membrane-bounded nucleus and organelles such as chloroplasts and mitochondria, themselves remnants of the ancient symbioses of several functional types of bacteria. Thus the eukaryotic cell served as an intermediate level of stability between the simple cells of bacteria and the multicellular organisms.

Step by step the holarchy has expanded in the "unbounded plans." Layer by layer, entities were constructed or evolved from the previous layer of wholes. These previous wholes became parts for the next layer, and so on, in a glorious effusion of stratified beings. It's a spiral process: Parts turn into wholes, which turn into parts for still more encompassing wholes.

Bronowski's concept points to a direction in the elaboration of systems in the cosmos. At discernible and successive levels, the rules of physics, chemistry, biology, and social psychology bring into play different kinds of relationships between the wholes, weaving them into systems that may in time be nudged into more and more definite wholes at another tier. In the simplest conceptualization of the spiral process, the emphasis rests squarely on how parts relating to other parts create the conditions that bring the wholes into being. Some systems theorists have termed this "upward causation"; it proceeds from the parts below to the wholes above. To avoid the temptation to say that things are getting higher, which is usually linked to the idea of getting better, perhaps we need to replace *upward* with another term. I also prefer to avoid the semantic troubles lurking in the word *causation* by replacing it with the less contentious word *influence*. A direction of influence radiating from a part to its more encompassing whole might thus be called an "outward influence."

A bicycle can serve as an example. A flat tire can put a halt to your whole trip. Thus the part not only affects but here determines the whole. This flow of influence—from parts to whole—can also be demonstrated in bicycle evolution. The invention of the wheel preceded the bicycle by thousands of years. Without quite so long a pedigree, so did chains, sprocket wheels, metal tubes, and so on. This is outward influence, a classic case of Bronowskian stratified stability in the evolution of complexity.

But what about the derailleur? The derailleur is the mechanism that switches gears by pushing the chain laterally across a series of different-sized sprockets and produces smudged fingers when the chain jams in between. This now-indispensable invention did not grace those first two-wheeled vehicles enjoyed by the pedalers of the late nineteenth century. The derailleur was developed after the bicycle itself. Here is a case in which the whole preceded one of its parts—in which the arrow of influence went the other way.

Technology offers myriad examples of this reverse direction in the arrow of influence between wholes and parts. Houses preceded glass windows. Cars preceded the microchips that have become essential to the headlights, door latches, and dashboard warning signals. In such cases the whole serves as the stable matrix within which new, better, more complex, more economical, or more whimsical parts are designed, installed, and tested.

Clothing and society constitute another excellent pair. Clothes are now essential parts of society. Most of us wear them every day. Imagine the streets of Manhattan on a hot summer day if everyone walked around naked. Yet society, we think, evolved before clothes. Society served as the matrix for the evolution of this now-indispensable part.

Influence directed from the whole to the part occurs throughout biology, too. Cells preceded multicellular organisms—true enough. But the earliest multicellular organisms had only a few cell types. The organism then served as the matrix for the generation of more and more types of cells. The types of parts became more complex in their numbers and functions within the stable embrace of the whole. Human liver cells, pancreatic cells, nerve cells, killer T-cells, and more than a hundred others did not evolve first as free-living protists in some pond and then magically come together into us. These cell types evolved within animal bodies as the bodies themselves evolved over hundreds of millions of years.

Always a crucial point, naturally, is the level of generality in the discussion. Cells as a type of life preceded multicellular organisms, but multicellular organisms as a type preceded particular, specialized kinds of cells. Parts can precede wholes and thus act on the coming into existence of those wholes. Wholes can precede parts and thus act on the coming into existence of those parts. If the first is outward influence, then the second may be called "inward influence."

We should therefore scrutinize Earth by looking for *both* directions of influence in the gaian holarchy. Clearly, nitrogen-fixing bacteria and rain forests affect the whole biosphere. We can serve up numerous examples of how other parts of Gaia also affect the whole in outward influence. But what about inward influence? Considering our bodies, we can argue that the inward influence on the organs and cell types was definitely tied to biological evolution itself, because the body is part of a larger system of evolution. This analysis cannot be analogously applied to effects from Gaia on its parts, because Gaia is not part of a system of evolution. But a different kind of inward influence may nonetheless be at work. We will thus want to look for ways in which Gaia is inwardly influencing its parts.

⟿ Gaian Cycling Ratios

Typically, we describe unusual objects in ways already familiar. To a child, a lion may be a big house cat. Many of us first learned about a platypus as a synthesis of duck and beaver. A laser might be thought of as a flashlight whose beam doesn't spread. The unusual degree of Gaia's singularity makes Gaia, too, difficult to talk about. So understandably, an initial, naive approach might be to call Gaia an organism outside the strictures of evolution.

Describing a laser as a nonspreading flashlight and Gaia as a non-evolved organism may be ways to launch one's thought when confronted with the unknown. But as with the laser, so with Gaia; easy initial descriptors all too quickly end up marooned on the shoals of oversimplication—or outright error. Also erroneous for Gaia is the machine analogy, because machines are parts of their own technological systems of evolution. One could play it safe and declare Gaia to be a system of interdependent parts and processes, a holarchy of many levels. But could that definition not also apply to a rock and its mineral structures? Organisms and machines are within systems of evolution. Rocks (and Gaia) are not. Fair enough. But likening Gaia to a rock doesn't help us off the shoals either. What next?

Gaia's singularity should be the focus. Ecologist Lee Klinger, specialist in the succession of bogs and a founding member of the Geophysiological Society, several years ago was talking with Connie Barlow, originator of the central insight of this section about the degree of closure of Gaia (we will talk about this in a few pages). Somewhat in exasperation at wrestling with endless webs of conundrums spun by all metaphoric answers to the question of what Gaia is, Lee finally insisted: "A cell is a cell, an organism is an organism, Gaia is Gaia."

Because of the natural and all-too-seductive tendency to describe Gaia in terms of known things or known systems, those who contemplate Gaia could do worse than to run in their minds, at least in the background, the mantra "Gaia is Gaia." The repetition prods one to keep the focus on Gaia's uniqueness, both as a holarchy and in a holarchy. Gaian mechanisms will probably reflect systems of logic different from those governing the organisms *or* rocks within.

The breathing of the biosphere, for example, as shown in the Mauna Loa record, differs fundamentally from that of our bodies. Oxygen gas is released as a by-product from the water split by photosynthesis. The hydrogen freed from the water is welded to carbon dioxide, forming an organic hydrocarbon. Within plants, to be sure, much carbon spins over and over in some amazingly complex metabolic cycles. But in these internal cycles, carbon remains in the generic family of hydrocarbons, as different types of organic molecules. We then eat the plants and spin that carbon in intricate internal cycles as well. Whenever that carbon, however, is oxidized to carbon dioxide, its eventual and ultimate fate, then its minutes on our home turf are numbered. We must purge it from our bodies through the lungs. We cannot perform the trick of returning it to the hydrocarbon realm. That happens at the larger level of the holarchy, the level of Gaia. The human body's fabulous, one-way pumping action brings oxygen in and forces carbon dioxide out. But Gaia does more than pump: It recycles. Gaia subsumes large fluxes of both sources and sinks of carbon dioxide and oxygen. The respirers and photosynthesizers—as two global organs or tissues—interconvert the air's molecules by ripping apart and recombining the constituent elements, as though one were digging a hole and the other filling it: no net change.

Were we human respirers the only life on planet Earth, carbon dioxide would rise and rise. And that's happening now, for in addition to the six billion of us, we have spawned about thirty billion technological equivalents of human metabolism, as measured by the release of carbon dioxide from fossil fuels. We nourish these energy slaves with the remains of ancient life. Oil, coal, and natural gas are yummy to the slaves, whose absolute numbers are growing faster than our own. In the outgassed breath of the energy slaves, we see what happens to an atmosphere subjected to a preponderance of respirers.

Any single type of organism, existing by itself, would throw Gaia out of balance. Like us, the filamentous mats of *Thioploca* generate CO_2. Like us, they oxidize their food. But unlike us, in this oxidation they consume nitrate. So in a world populated by these microbial spaghetti strands alone, the nitrate in the ocean would drop dramatically. All organisms have metabolisms that by themselves would throw everything out of kilter. A monoculture planet is therefore a thought experiment with no place in reality.

Consider: Toiling below *Thioploca*, the sulfate reducers produce hydrogen sulfide and deplete sulfate. These flows move in opposition, and thus are potentially complementary, to those of *Thioploca*. At the very least, it is known that the source of hydrogen sulfide from the lower beasts becomes coupled to the sink for that compound in *Thioploca*. An ecosystem is born.

One broad division in ecosystem research focuses on nutrient cycles. Such research considers how the wastes from all the members of the food webs travel into the masses of fungi and bacteria to complete the nutrient cycles that ultimately resupply the photosynthesizers. Ecologists Tim Allen and Thomas Hoeskstra see nutrient cycles as the very markers for designating the existence of ecosystems. Ecosystems don't evince borders as visual and tactile as our skins. Their borders are rather defined functionally, as places where fluxes to or from the outside become small relative to the flows within the interior cycles. The relative closure of the cycles provides such entities with definition in space, at least to minds that demand delineation and a name.

Looking into the details of such a functional definition of an ecosystem, however, typically renders borders fuzzier than the fog of San Francisco. Any organism participates in a phenomenal range of different-size cycles. A tree, for example, takes part in very local cycles in the soil at its roots and in very distant cycles via the birds that consume insects that feed on its foliage, especially if the birds seasonally migrate a thousand miles. The inevitable fuzziness of such borders fuels seemingly endless debates in ecology about just how connected the parts of an ecosystem are and about just how discrete ecosystems are as entities.

Take Pine Woods Park, for instance. Nutrient cycles aplenty roll within it—for example, the nitrogen and phosphorus cycles. But how long could the whole sustain itself if not for open borders allowing exchanges over, under, and across the surrounding streets? Migrating birds must move in and out with the seasonal shifts in food supply; ground water must flow. More to the point, the park needs to exchange gases with the global atmosphere.

What if we admit the air—but only that directly above the park—

as part of the system? Perform a thought experiment: Bottle the system by erecting vertical glass walls around the park, eight miles high to the top of the troposphere. What will this mean for carbon dioxide?

Today Pine Woods Park is returning to its old glory. Because some original trees have survived, the regrowth is not as intense as it would be in a plot making the transition from abandoned farmland all the way back to forest. But with the typical photosynthesis of a temperate deciduous woods, the park could easily be toting up an annual net increase of between 100 and 200 grams of carbon per square meter, stored in the form of biomass, both above ground and in litter and soil. How much carbon is in the air above it as CO_2? The eight-mile-high column of air above each square meter contains about 1500 grams of that element. Given the glass borders, that doesn't leave many years before starvation from lack of gaseous feedstock sets in. This is not a calculation in a fantasy world of photosynthesizers without respirers. This is a real situation (except for the glass walls), an ecosystem with both photosynthesizers and respirers running at their own level of balance or imbalance.

OK, you say the situation is unnatural because the park had been cut back for many years by the townsfolk who are now allowing it to regrow. But even more dramatic situations occur in "natural nature." Fires in the western United States can devastate areas equal to a thousand Pine Woods Parks. These areas can regrow only because they are open to a globe-spanning atmosphere. OK, now you say that the carbon dioxide generated from the fire would have gone into the hypothetical bottle above the charred area and thus would be available to supply the regrowth. In that case, the gas could be going up and down by a factor of two to four. That's certainly not how the biosphere works now. Besides, what about the rain, if such small areas were isolated from the open exchange of gases? What about the monsoon?

For life, stability in the sum of vital fluxes ultimately rests on cycles at the global scale. A way to explore this quantitatively is to analyze some particular cycles in relation to their entering and exiting fluxes. For the sake of simplicity, let's assume that these fluxes into and out from the cycle itself are equal. Also for simplicity, let's ignore human jostling of any cycles.

In the ocean's surface photic zone, carbon atoms shunt among a teeming ecosystem of algae, small zooplankton, and bacteria. Carbon moves into and out of different biological forms and its ionic forms in the water. The cycle is not perfect, however. Carbon is lost because detritus drops out by gravity and organic molecules are mixed downward by turbulence. A fish feasting at the surface and then swimming 400 feet deep to excrete waste contributes to the loss as well. On a global average, about ninety percent of the carbon taken up in photosynthesis by the algae and cyanobacteria eventually returns in a local loop that allows them to have another go at it. The remaining ten percent exits from the surface system.

We might define a cycling ratio as the ratio between the flux passing through the photosynthesizers within any given cycle and the flux exiting from the cycle. In the ocean's photic zone, the cycling ratio compares the cyclic flow of carbon within the surface system and the exiting loss of detrital flux downward. The ratio is 10 to 1. This means that on average, an atom of carbon will circulate among algae, zooplankton, bacteria, and bicarbonate in the photic zone about nine times, and then, after the tenth trip through the algae, it will exit, stage down.

Let's look at some numbers. Marine photosynthesis incorporates about 40 billion tons of carbon each year into biomass. One-tenth of this amount is 4 billion tons, an amount that leaves the surface layer of the global ocean as detritus. How are these 4 billion tons resupplied to the surface? In computing the surface cycling ratio, I ignored gas exchange with the atmosphere in order to draw attention to the organic loss of detritus from the surface ecosystem. But in asking questions about resupply of losses, we must consider the atmosphere because it is intimately connected with the chemistry of the ocean. The amount of carbon in the atmosphere's carbon dioxide is too little to be a long-term panacea for any problem of carbon losses. Nonetheless, because the ocean and atmosphere exchange gases with each other, we should probably enlarge our compass of computation and include not only the ocean

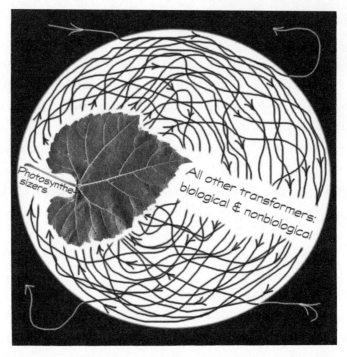

A cycling ratio. Here the pathways enter and leave the photosynthesizers twenty times as an element is passed around all other transformers, both living and nonliving, within the system of the white circle. Across the border between white and black, arrows enter and leave only twice. This case visually demonstrates a cycling ratio of 20/2 = 10.

and atmosphere, but also the terrestrial biota and soils — in other words, the entire gaian system.

What is the supply flux of carbon into the gaian system? Supply comes from several sources: from the dissolution of rocks such as whitish limestones that contain calcium carbonate, from the oxidation of fossil organic carbon in rocks such as shales, and from the fumes of volcanoes. All together, these sources provide about a half-billion tons of carbon per year.

Note that this half-billion tons of annual input pales before the requirement of 40 billion tons per year for marine photosynthesis. The input to Gaia cannot even come close to resupplying the 4 billion tons of carbon annually lost from the surface layer of the ocean. This vividly demonstrates what is indeed known to be true: The 4 billion tons of detritus from the ocean's surface may be a loss relative to that surface, but it cannot possibly be an ultimate loss to the gaian system. Otherwise, the total amount of carbon would be plummeting, and the system would be going badly out of whack.

Much of the detrital loss from the ocean's surface is in fact recycled within the ocean. Although the waste that fish deposit and other, finer organic matter in the detrital flux fall out of the frame of analysis drawn around the photic layer, the carbon has not escaped the widely dispersed clutches of life. If the frame's lower border is allowed to descend, it now encloses other living things, both large and small. Vertically mobile creatures and the debris-attached bacteria in the deep waters respire carbon dioxide from the organic, detrital carbon. The regenerated carbon, in the form of water-borne bicarbonate ions, travels upward and is eventually returned to the surface by ocean turbulence. In the steady state, which must be close to what we have in the ocean at all times, the bicarbonate ions in deep water diffuse upward and exactly replace the carbon lost as detritus from the surface.

Even for the small fraction of detritus that makes it through the water column's hungry sieve of life to settle at the bottom, hundreds of feet or miles down as the case may be, some of the carbon will still be recycled by worms and bottom-dwelling bacteria such as *Thioploca*. How much, in the end, is buried beneath the reach of the hungry? In real numbers, about a half-billion tons of carbon per year exits the system, entombed in the sediments, mostly as calcium carbonate shells but also as organic debris, destined to join the next round of limestones and shales in the geological cycle of uplift.

How does life know enough to bury about the same amount of carbon as that which enters the gaian system? It doesn't, of course. There is no easy general explanation, unfortunately, for this balance between inputs and outputs on the global level, for most elements over long enough time scales. But consider an analogy. Imagine a bucket supplied by water from a faucet. The bucket has a small hole at the bottom. The water that runs out of the hole exactly balances the input from the faucet, because the water will adjust to whatever level is required to make the exiting flux equal to the entering one. If the hole is small, the water level will be high in the steady state. If the hole is larger, the water level will be lower. This analogy is simplified to a degree that is almost embarrassing, given that it must represent the entire complex system of chemical reactions and all ecological tugs within Gaia that adjust exiting fluxes to the entering ones. But this analogy is crudely applicable for the chemical elements within Gaia.

Global photosynthesis currently uses 60 billion tons of carbon on land and 40 billion tons in the ocean each year, for a total of 100 billion tons of carbon. Yet it lets slip through only a half-billion tons (basically all in the marine sediments). The global—or gaian—cycling ratio for carbon is thus about 200 (100/0.5). The exchanges within Gaia allow an atom of carbon to circulate 199 times into photosynthesizers, out again to respirers and the ocean and the atmosphere in a myriad of forms and places, and then back again to photosynthesizers. Finally, on the two-hundredth trip on average, the carbon atom leaves via burial.

One can invert the logic of the cycling ratio to gain another perspective on what the number means. How much photosynthesis could occur if plants and algae had to rely solely on the supply flux of carbon from rocks and volcanoes? About a half-billion tons of carbon per year. If photosynthesizers were denied access to recycled carbon from their own productions, global photosynthesis would be limited to half a percent (0.5/100) of what it is today. This viewpoint reveals the cycling ratio as a measure of the amplification bestowed on life by the interior dynamics of the gaian system, born from both the living and the nonliving. Recycling within the frame of Gaia, with respect to carbon, amplifies the photosynthetic flux 200-fold.

A similar analysis can be applied to nitrogen. Global photosynthesis takes in between 7 and 8 billion tons of nitrogen per year. Estimates from the dissolution of rocks and in volcanic gases are that about 6 *million* tons of nitrogen per year enter Gaia. The amount buried in the ocean as organic detritus from life is about 14 million tons per year. Note that this burial is not exactly the value of the entry flux, as in the bucket analogy; but compared to the amount going into photosynthesis, these estimates for entry and burial fluxes must be considered to be remarkably close and to confirm the essential balance. Use either 7 or 8 billion tons for the photosynthetic requirements, and, for the denominator of the cycling ratio, use either the supply flux or the burial flux. This yields a range of between 500 to 1300 for the gaian cycling ratio for nitrogen. That's a lot of amplification.

A cycling ratio for nitrogen might be thought meaningless, because the atmosphere is such an immense pool for nitrogen. Is it so immense? The atmosphere contains 4 *million billion* tons of nitrogen as N_2 gas. That does seem immense. But if this pool had to supply photosynthesis with 8 billion tons per year, and if the products of photosynthesis (nitrogen only, somehow) were removed and never recycled in any form, even from rocks or deep volcanoes, then the atmosphere's nitrogen would be used up in only half a million years. Certainly, deep ocean volcanism would recycle some of the biological nitrogen to nitrogen gas (as it presumably would recycle nitrate in the absence of denitrifiers) and thus prevent this catastrophe. To some unknown extent, that is. But the point is that this is not how the system works now. It works by vast and efficient recycling *within* the gaian system.

These high values for the cycling ratios of carbon and nitrogen on the level of Gaia bring to the fore the unique property of Gaia. It is difficult to compare Gaia directly with organisms in degree of recycling, because although organisms perform a lot of recycling, they keep their elements in certain forms—for example, carbon within organic molecules. The key to understanding Gaia as an entity is the fact that many chemical transformations needed to cycle the crucial elements occur among different types of organisms, not within organisms. The sulfate reducers in the black muck produce hydrogen sulfide from sulfate; *Thioploca* makes sulfate from hydrogen sulfide. Photosynthesizers create organic carbon from carbon dioxide; respirers use organic carbon and return carbon dioxide.

The cycling ratio within the surface photic zone of the ocean is only about 10, compared to the gaian cycling ratio for carbon of 200.

Another cycling ratio internal to Gaia can be formed by comparing the nitrogen needed by land plants to the exiting flux of denitrification from terrestrial soils (balanced primarily by the entering flux of nitrogen fixation). This ratio is about 12. Compare that to the gaian cycling ratio for nitrogen of between 500 and 1300. Thus many of what are paradigm systems of recycling on a less than global scale (the ocean's surface layer for carbon, the land's soils for nitrogen) are inefficient compared to the degree of cycling—and thus amplification—provided by the entire, integrated system of life, atmosphere, soil, and ocean. In the holarchy of life, Gaia is more closed than any of its subsystems, which, by comparison, seem more like flow-through systems. If entities are defined by borders, Gaia qualifies. It does not have a skin, membrane, or wall. Its border, rather, is functionally defined by the relatively small mass fluxes that cross between inside and outside, compared to the massive cycles among interacting life and the fertile chemical baths of solids, liquids, and gases.

The small mass fluxes for carbon and nitrogen from rocks and volcanoes are a reality, a reality that life has had to work with and evolve within. The fact that small mass fluxes define a border to Gaia and create conditions that foster the high cycling ratios inside is an example of inward causation. Crucial to the development of high cycling ratios for some elements (but not all, as we will see later) is the fact that the amount of sunlight falling on the planet makes the potential for incorporating carbon and nitrogen on a global scale much, much greater than the supply of these elements from beneath Gaia. As we will discover in a later chapter, photosynthesis could be even higher. Thus there is no doubt of the potential. For example, the input of half a billion tons of carbon into Gaia each year works out to a gram of carbon per square meter over Earth's entire surface. Regrowth in Pine Woods Park accesses a hundred times that.

The near-closure of the material cycles at the level of Gaia could float us off the shoals of confusion about the nature of Gaia. The remarkably tight seal of matter on the global scale acts as an inward influence, from whole to parts. It is surely not so active an influence as the demands of natural selection pressing on organisms everywhere. The closure of matter on the Gaian scale is more passive, a constraint imposed on the parts, but it may have been vitally significant to the evolution of the intricate scales of life inside.

The situation somewhat reminds me of the stretchers of a canvas to be painted. The framework bounds the arena for the creativity about to unfold. So the small mass fluxes at the borders of Gaia act as boundary conditions for the creativity of evolution inside. For the biosphere, a constraint becomes a glorious opportunity. The fact of small material fluxes across its largest, defining borders reveals the uniqueness of Gaia, a property of the whole impressed on the parts, the secret of understanding Gaia's body.

3
Outer Light, Inner Fire

Southward along a line of longitude our research ship steamed, halfway between Africa and South America. That day we were to cross a line supposedly visible only on maps. A ceremony would be in order for all those making the crossing for the first time, including me. Certificates would be handed out several days later, when a lull in the busy schedule allowed all on board to gather as a group. But before celebration comes initiation.

In the torrid heat of morning, the half-dozen initiates were bundled into thick yellow rainsuits and assigned mind-numbing tasks. Mine was to clean the ship's bell. I sweated and slaved with a weak brush and still weaker cleaning solution to remove the encrusted crud molecule by molecule.

I abandoned the bell temporarily for the lab, however, when the ship stopped in the middle of blue nowhere. Throughout this trip, it was my responsibility to measure radon concentrations in water samples from various depths at pre-set locations. Still sweating in the clumsy rainsuit, I cursed about tax dollars for these experiments being jeopardized by such an idiotic ritual. There was some consolation in remembering that a hundred years earlier I might have been roped and tossed overboard into the ship's wake and dragged through the invisible line.

The real reward in crossing the equator was not a certificate. It was experiencing the equator as an actual presence. Somewhere near the theoretical line, the winds from the north and those from the south converge. The zone shifts with the seasons, and like any climatic feature, it is patchy in space and time. This unpredictability will discourage tour ships from ever heading equatorward just to witness the convergence. But on satellite maps averaged over a sufficiently long time, it emerges as one of the truly great phenomena of world climate. It shows up as a shiny ring around the planet, a white girdle of clouds against the dark ocean water. Climatologists call it the intertropical convergence zone, the ITCZ.

Winds are pulled toward the equator for several thousand miles from both hemispheres. The phenomenon is somewhat like the summer monsoon in the American southwest, which draws air inward to replace the rising air over the continent. The equator-bound winds stream in from both sides, increasingly laden with vapor swept up from the sea. Finally, at the convergence, the winds switch directions from horizontal to vertical. There the hot tropical air rises with vigor. Cooled in the rising, the moisture condenses. Clouds are born in a globe-girdling ring of white, the nearest thing to an east–west line that you'll find in nature.

This ring of moisture circles slowly westward and undulates even more slowly north and south, following the solar seasons and releasing its cargo over any land masses it encounters. We see the results as a broad zone of green: the rain forests of the Amazon, the Congo, Southeast Asia, and Indonesia. Injected into the turbulent weather patterns over the continents, the air masses born in the ITCZ can spread like soft butter. But sometimes the beclouded convergence is narrow, its buoyant markers of clouds patchy and unpredictable. On average, it is lodged north of the equator, an asymmetry caused by the larger land fraction in the Northern Hemisphere. In winter the ITCZ moves quite close to the equator, when and where I was fortunate to be.

In the hour before darkness I stood virtually alone on the deck, near the accursed but now shiny bell. Everyone else was dining below. Why aren't they all skipping the meal to admire the skies? The spectacle was unlike anything I have encountered before or since. It was like hovering in a glass helicopter right amid the clouds. And simultaneously, it was like peering up from a valley into clouds far above. An extravagance of cumulus clouds of all shapes and sizes surrounded me. Mixed in like streaks of paint were flatter clouds of every shade of gray. Nearly in orbit, glowing traces of cirrus clouds embroidered the furthest reaches. The sensation was not of overcast but of openness, a latticework of clouds and plentiful pockets of blue space. The slashes of horizontals and the billowing verticals created a hypnotic beat in my mind, two alternating chords of a meteorological symphony.

The tropics are famed for their biodiversity, from plants and beetles to bats and ants. From the ship, I could visit this exuberant flowering of species only in my imagination. To the distant east was the African Congo, Conrad's heart of darkness, but dark only on the old maps of Europeans. In reality it is a lush, fat, verdant thumb of biodiversity tucked into the palm of the continent. To the distant west was the Amazon, the watershed of the world's mightiest river, where a single tree harbors more species of insects than most of us could learn to recognize in a lifetime.

In my all-too-brief passage through the convergence zone, I witnessed perhaps the world's most magnificent diversity of clouds. The geographical equator is a line on a map, invisible at sea, a joke except for those earning their silly certificates. But the climatological and biological equators are real — not so much things as presences. Biodiversity in the tropics, cloud diversity in the tropics: Might they be related by the energy from the sun?

The World's Whirls

The ancient names for the sun—goddess Amaterasu of Japan, god Ra of Egypt, god Helios of Greece—show that people everywhere have honored the sun as a special entity in the cosmos. What would happen if the sun were to turn off for a week? Just extrapolate from the cooling that descends over a single night as the atmosphere's heat radiates to space. The feeble heat from Earth's interior would barely keep the surface above absolute zero. To understand Gaia, we must first understand this planet's relationship with Helios.

For Gaia, the most important fact about the sun, besides its stupendous incoming energy, is the geometry of its rays. About ninety percent of the rays that illuminate our surroundings travel in nearly perfect, parallel alignment. Typically, ten percent of the light falling upon Earth's surface is diffuse, scattered in all directions by the sky's molecules. But this ten percent is crucial: It makes the sky blue. At the upper border of Gaia—the top of the atmosphere—all light is direct, the rays virtually parallel. The maximum possible deviation here is in the rays that converge to a single observer from any two opposite edge points of the solar disk: about one-half a degree off perfectly parallel.

Like the sun, Earth is a nearly perfect figure from Euclid's geometry: a sphere. It is well known, but impossible to experience directly without tedious measurements, that Earth flattens at the poles and bulges at the equator. It misses the geometric ideal by about 13 miles out of an 8000-mile diameter. You'd be hard pressed to buy a plastic ball at Wal-Mart that was as smoothly spherical as Earth.

Parallel lines of light traveling through space dash themselves upon a rather dark, solid sphere, marbled with spirals of white that reflect about 30 percent of the light. This circumstance determines Gaia's fundamental geometry, with broad consequences for all levels of the holarchy.

Étienne-Louis Bouleé, a visionary French architect living at the time of the French Revolution, proclaimed the sphere the most beautiful shape of all. When the sun falls upon a dome, he said, we delight in the infinity of different shades. On Earth's dome, the brightest spot corresponds to the place where the sun is overhead at any given moment. Around this brightest point are concentric circular zones, of whatever fine gradation one cares to note, where the sun's intensity fades into ever darker tones. This progressive darkening is due to the decreasing angle between the plane of the local surface and the arriving rays of light. Shallower angles spread the light over larger areas. For a stone dome in sunlight, this means a gradient of illumination. For Earth, it means a gradient of primary heating.

Imagine a sphere floating before you and illuminated by light in a perfectly parallel beam. The light comes from you. Add, through the sphere's center, an axis tilted so it would pass through the marks on a clock at 4 and 34 minutes. This approximates the Earth's 23.5° tilt from North Pole to South Pole. The equator (adding it in your imagination) stretches between the clock's marks at 19 and 49 minutes. Start the Earth spinning from west to east about the tilt axis. Choose a favorite city at some latitude, Florence at 44° north, St. Petersburg at 60° north.

Sunlight falling on tiny spheres. The situation is like that of Earth, a larger sphere.

Follow the daily rise and fall of illumination on the city as the sphere spins. This position in the visualization corresponds to a permanent spring equinox on Earth as seen from the sun.

Now run a second axis directly vertical, through the center of the sphere, passing between the marks of 12 and 6 on a superimposed clock face. To see how the Earth looks during the course of a year from the point of view of the sun, it is necessary to rotate the already spinning sphere about this vertical axis at a slow pace, by gradually swinging the tilt axis from the right to the left toward you until the tilt appears to be vertical relative to your line of sight — the summer solstice. To be precise, you would have to put just over 91 spins around the tilt axis during this quarter of an annual turn about the vertical axis. Continue the process. It is similar to how a top spins rapidly while simultaneously turning around a longer, wobble cycle. Pick out your favorite city again and follow it during its daily cycle, noticing, furthermore, how the daily sequence of illumination changes over the year. If you can do all this without getting mixed up or inducing a headache, you are a far better visualizer than I am. It is a difficult but rewarding exercise.

Fairly easily, however, one can visualize that during the year, the greatest abundance of direct light falls, on average, upon the equator. Sites on the equator daily spend more of the time in the brightest zone of the sphere than do sites at any other latitude. It is also easy to visualize how the poles pass into light and then into darkness not daily but only on the annual cycle. The complex patterns that the parallel shafts of sunlight make upon this doubly rotating sphere describe the daily and seasonal patterns of energy input to Gaia.

The relevant quantity is the energy flux absorbed by the surface of land or water and by the vertical column of air all the way to the top of the atmosphere. How is this total absorption distributed as a function of latitude? From about 20° south to 20° north latitudes, absorption of energy is fairly uniform, which is why everywhere in the tropics is so similarly, well, tropical. Poleward in both hemispheres, the values plummet more and more steeply with increasing latitude. At New York City's latitude near 40° north, total annual absorption is less than 70 percent of the tropical value. The absorbed annual bounty from Helios around 60° north, say St. Petersburg, is less than 45 percent of the tropical gain. From equator to pole, across the sphere a primary gradient of energy produces a gradient of temperature.

Molecules of hot air move faster than molecules of cold air. Wherever warm and cold are in contact, the difference drives a transport of energy. We see and hear this transport as wind, which rustles leaves, pushes waves, and sweeps clouds across the sky. Wind is caused by pressure differences, which in turn are primarily caused by temperature differences along Earth's surface and in its air. On the grandest scale of the planet, the dominant pattern of differential heating originates from the parallel shafts of sun falling on the sphere.

At the intertropical convergence zone, the warm, buoyant air ascends into the spectacular clouds that I witnessed. The rising air cannot descend back where it came from because more air is rising behind it. Thus at the top of its draft, it gushes laterally both north and south, a high-altitude divergence complementing the convergence at the surface. A convection loop is thus established, an atmospheric cell. When the air that rose out of the ITCZ finally descends at the other side of the cell, it is thousands of miles away and dry as a bone. This descending curtain of dry air is responsible for some of the deserts in belts around 30° north and 30° south latitudes. Australia, for example, sits in one path of descent. So does the Sahara.

The ascending ring of the ITCZ and the northern and southern descending portions a third of the way to the poles form two circulating doughnuts around Earth—the famous Hadley cells. The planet possesses other giant, circulating cells, such as the Ferrel cells in the midlatitudes, the polar cells, and the east–west Walker cell in the Pacific. But these are only the biggest spins. A universal property of turbulence is a vast range of scales of motion, from gargantuan to minute (seen, for example, in the eddies in a river or in an ascending stream of smoke). You can also experience these scales of spin when the wind shifts direction against your face while overhead clouds march in unwavering formation across the sky, a small portion of a state-spanning

Four days of white whirls. These views from above the Pacific Ocean, on the left, and from above North and South America, on the right, progress downward, from October 23 to 26, 1996. In the image at the top right, a hurricane, which has spun off the American coast and is about halfway across the Atlantic Ocean, can be seen to dissipate over the days. Several of the pictures show the white, horizontal band of the ITCZ, just north of the geometric equator in the Pacific Ocean. Images provided by NOAA/GOES and SSEC, UW-Madison.

mid-latitude gyre. Lewis Richardson, English physicist and meteorologist of the early twentieth century, wrote this couplet on the scaling of turbulence:

Big whirls have little whirls that feed on their velocity.
Little whirls have lesser whirls and so on to viscosity.

The ocean joins in the turbulent play. First of all, its fluids are brushed around by the winds. Just as one might start a floating inner tube spinning by pushing one side with the left hand and pulling the other side with the right, the easterly winds in the tropics and westerlies in the mid-latitudes combine to propel gigantic surface gyres in the ocean basins, clockwise in the north, counterclockwise in the south. In the north these gyres intensify along the eastern coasts of continents, producing swift and wide currents out at sea: the Gulf Stream in the Atlantic and the Kuroshio Current to the east of Japan — flows that put the Amazon and the Ganges to shame.

But the ocean is more than a pushover to the wind. It forms its own unique kind of convection cell: the thermohaline circulation. In contrasting symmetry to the way the atmosphere is driven upward by heating at the equator, the ocean is driven downward by cooling near the poles. *Thermo* means heat, *haline* refers to salt: At particular places in the polar waters, such as in the North Atlantic and around Antarctica, combinations of these two factors initiate a deep circulation that encompasses the entire world ocean. Wintertime cooling makes the water denser. So do evaporation and the freezing out of fresh water, which strand the salt in the remaining surface water. At sufficient density, the polar waters sink all the dark way to the deepest benthos. There they spread out, flooding the deep layers of the sea with cold.

The day before my equatorial crossing, I decided to verify what the books had said. After the other scientists had drawn their allotments of water from the rosette of sealed canisters, they opened the sampling spigots so the canisters could drain. I reached under the canister that held a sample of the deepest equatorial water hoisted from three miles beneath the ship. It was ice cold. Several hundred years ago, that water may have sunk in the Weddell Sea off Antarctica, five thousand miles away. As the deep water circulates, portions of it peel upward to form an ocean-wide convection cell that broadly upwells at the less-than-snail's pace of about fifteen feet per year. Eventually, this water, near the surface, returns toward the poles. This overall thermohaline circulation takes on average a thousand years for just one complete revolution—another physical cycle ultimately called into being by the way light falls on a sphere.

When you sit in a bathtub with the hot water running, very soon your feet get too hot and your butt too cold. What to do? Start stirring with your feet. Adding the hands will even out the temperature even faster. Turbulence effectively moves heat.

Whirls of air and water with rising and falling portions of cells, with horizontal pancake spins, some the size of real pancakes and some gaian-scale pancakes: All result from Gaia getting too hot at the equator and too cold at the poles. Planetary feet are kicked and hands swished by the automatic responses of thermodynamics and fluid mechanics that move heat. One directive for gaian inquiry suggested that we attend to the cycles of matter. Here some are.

What is the impact of the whirls of gas and liquid on the planetary physiology? We think of the tropics as sultry and of the poles as frigid. True enough. The average air temperature over the year at the equator is about 80°F, above water a few degrees warmer. The coldest waters in the highest latitudes of both hemispheres are near 32°F. (Because of the salt, ocean water can drop a few degrees below 32°F without freezing, and it often does, particularly around Antarctica.) Rounded off, the temperature difference is 50°F between equatorial and polar waters. The actual frozen poles are colder still. Let's stick with the Northern Hemisphere; the South Pole's temperature is extra chilly because of the high altitude. The North Pole's annual average air temperature above its floating ice sheet is about −10°F, making the equator-to-pole difference 90°F.

Because heat is continually pumped from low to high latitudes, we can assume that if there were no such pumping, the tropics would be even hotter and the poles colder. But how much?

Imagine the whole world partitioned into plots bordered by high glass walls, like those we earlier imagined around Pine Woods Park, and extend those walls down into the oceans as necessary. This would block all large-scale aerial and aqueous circulations. Ignoring the small amounts of heat that might pass by conduction across the walls between cells, the total mass within each cell — the plants, soils, waters, and air — would reach a temperature that ensured a local energy balance between the radiations received from and sent to outer space. The received radiation is the solar flux, which, as we have said, depends on latitude because of the parallel rays falling upon the sphere. For the purposes of the thought experiment, assume that clouds remain the same; they cover about 50 percent of Earth and reflect about 30 percent of the light, fairly uniformly). The radiation sent to space consists of the infrared rays emitted by the matter within the plot, the amount of which depends, all else being equal, on the surface temperature. Given that the grid of glass walls checked any lateral circulation, within every plot of the grid, the incoming solar radiation would be balanced by the outgoing infrared radiation, and the local temperature would be set by that balance.

Satellites can directly measure the infrared flux to space at different points above the surface. The data show that in the real world, the infrared outflux at the tropics is less than the influx of solar radiation. Of the 300 watts absorbed by each square meter of typical tropical ground, only about 240 watts make it back out to space from that surface. The missing 60 watts per square meter must go somewhere. It is transported toward the higher latitudes, both north and south, by the great swirls of air and water. Thus in the high latitudes, the situation is reversed: More energy exits as infrared radiation than enters as solar influx, and the numbers are even more dramatic. The poles receive a paltry 40 watts per square meter from the sun but radiate 160 watts per square meter as infrared to space. The additional amount of 120 watts per square meter has been transported poleward by the turbulence.

So how sultry and how frigid would the extremes of Earth be without this lateral heat transport? In current climate theory, a change in energy of a watt per square meter would produce a temperature change of roughly 1°F. Thus in a world without the whirls, instead of today's 80°F, the tropics would climb to 140°F—unbearable for plants, animals, and fungi as we know them, though exotic, heat-loving bacteria would thrive. Instead of today's −10°F, the North Pole would be −150°F—bad news for Canadian polar bears. Recall that this calculation uses the annual average solar energy. In reality, without the whirls, the poles during their six months of darkness would drop to nearly absolute zero (−476°F), in thermal equilibrium with deep space. During their summers, with 24 hours of sun, the poles would reach 50°F, for an annual swing up and down of more than 500°F.

Without the whirls, the temperature difference between tropics and high latitudes would be magnified by 3 to 4 times. Whew! Compared to what life is used to, the planet would be inhospitable. In addition, the ocean would turn to ice all the way down wherever the surface temperature fell below the freezing point of brine. Instead of the cold stream that I felt from the canister, the equatorial deep water would be the same 140°F as the tropical surface. At one latitude, though, conditions at the surface could be about the same as today. At about 30° north and 30° south latitudes, the energy received from the sun and that emitted spaceward by the surface are currently balanced. It would surely be a weird world without the heat pump, a planetary mixer that exceeds our technology's total power output by a thousand times.

We can say that organisms obtain a variety of climatic environments from the unequal distribution of sunlight. But paradoxically, life also gains a measure of climatic equality, because the skewed distribution of sunlight causes the very whirls that redistribute the heat and level out the temperature. Life inhabits a spherical shell with dynamics that produce a state between sameness and extreme difference.

Life gains much more. What could be better than getting a global circulatory system for free? The seasonal breathing of the terrestrial biota that we measure in the air above Mauna Loa in the middle of the

Pacific Ocean shows how rapidly the atmosphere mixes, particularly along belts of latitude.

The intertropical convergence zone, it turns out, presents the only substantial barrier within the whole atmosphere. As the converging winds from both sides rise in the vicinity of the equator and then diverge at a higher altitude, parcels of air tend to keep to their own sides and not sneak across to the other hemisphere. The biggest leak between the two hemispheres, on the other hand, is impelled by the giant of monsoons. When the summer sun heats the vast Tibetan Plateau, pulling in marine air (which refreshingly drenches India), the suction is so powerful that the replacement air is dragged all the way from the Southern Hemisphere. Overall, and even accounting for the barrier of the ITCZ, the global mixing time is only about a year. Thus all organisms, from the crabs at the southern tip of Tierra del Fuego to the reindeer of Siberia, are separated by no more than a year in shared breaths.

Life in the ocean, too, shares pretty much the same water because of the free global circulation. The amounts of the big ions—chloride, sodium, sulfate, and magnesium (which occur in that order of abundance)—generally vary by at most ten percent over the bulk of the world's ocean basins, and by at most twenty percent when certain regions of the oddball Arctic Ocean are included. That tiny ocean is odd because bottlenecks in the Norwegian Sea and Bering Strait limit its exchange of water with the Atlantic and Pacific. The Arctic's unique internal chemistry makes it the exception that proves the rule. Without the scales of spin in the sea, portions of the ocean would chemically vary by big factors, as do the relatively isolated seas: the Mediterranean Sea, the Black Sea, the Dead Sea. Deep water in particular is relatively uniform even in trace nutrients, because the thermohaline circulation spreads the dense, cold water from the polar seas all around the deep parts of the world. By contrast, the trace nutrients can differ significantly across the surface because of the activities of life and differences in upwelling.

Organisms generally control their interior circulation. Such circulation—our blood, the fluids in plants, the spinning cytoplasm in cells—helps unify the organism into an operational whole. Roots and leaves of plants share two fluids, one moving up, the other down, into and from which the plant cells give and take ions and molecules. Our blood circulates along tubes of different sizes, but also within the tissues, among all cells. These circulations were shaped by the honing forces that act upon organisms as parts of systems of evolution.

To life, the air and water of Earth are the extraorganismal fluids. Gaia, being singular, could not evolve a circulation for these fluids. But because globe-scaling transport systems faithfully swished across the planet from its earliest days, Gaia's circulatory system came for free.

This free circulation should be regarded as a given that affected life within Gaia, a case of inward influence like that of the closure of matter set by the limited fluxes across the biosphere's borders. Of course, the atmosphere was not the same at life's beginning, but the key is the mixing that served (and serves) as a common habitat for life. We eventually have to ask what effect this free circulation had on the global integration of life over nearly four billion years—on the organisms that evolved within a single, swirling atmosphere.

Chemical pockets of variety occur, of course, and moreover, they are vital. An example is the low-oxygen zone inhabited by *Thioploca*. But the segregation of environments implied by the chemical differences is always only partial. *Thioploca* would die without the organic detritus supplied by a surface life above its mats that is in contact with the same global atmosphere that you and I breathe. All organisms either breathe the global atmosphere or are in intimate, life-dependent contact with organisms that do. Thus ultimately, except perhaps for the newly dis-covered microbes miles underground, all life on our planet is connected via one blanket of air, one bath of water.

⌇ HEAT OF VULCAN

Cliffs the color of sandpaper tinged with orange surround me in the mountains of the American southwest. The rock is patched with lichen species distinguishable by their rainbow colors: lemon yellow, sunset orange, brown, gray, and a brilliant chartreuse. Accessory pigments mask the common green pigment of photosynthesis in the algal partner of the symbiosis. The fungus, the other partner in this classic biological binary, provides the actual attachment to the rock. In a lichen, a photosynthesizer and a respirer have physically united. This world in miniature is not necessarily balanced in its cycle of gases, however, and so, like all other systems within Gaia, it requires the global metabolism.

We all live between two worlds, earth and sky. A lichen embodies such a life more clearly than any other organism: Some are so thin and adhere so tightly to the rock that a probing fingernail can discern no thickness to their bodies. Even the crusty ones that can be peeled away lie flatter on their host rocks than the thinnest pancake on a sizzling griddle. With a body that is little more than a surface, a lichen bridges two environments.

Gaia, too, is a thin sheet between two vast environments. Like a soap bubble, only relatively thinner, Gaia possesses two surfaces, which might be called its dorsal and ventral sides. But unlike the soap bubble's two environments, which are basically the same (air at slightly different pressures), the two worlds across Gaia's borders are worlds apart. Gaia's dorsal surface faces space, black and empty except for night's candles, one of which is close enough to flush forth the day. Gaia's ventral surface presses against solid rock, a perpetual darkness relieved only by the red glow of lava or the eerie luminescence of deep sea vents.

On the continents, the solid rock at the ventral side of Gaia—bedrock—may be cloaked by a layer of soil. But where mountains are dry, bedrock is everywhere exposed. Consider a yellow lichen clinging to a tan bluff of bedrock. Is the base of the lichen identical to the base of Gaia in that tiny fraction of the biosphere's reach?

To answer that question we need to explore flux rates across surface boundaries. In our bodies, nearly two hundred liters of fluid pass each day through the cleansing glomeruli of the kidneys. We typically take in, via food and drink (and thus equally excrete), about two to three liters of fluid per day, although this can rise to six liters or more with heavy exercise. The flow rate of our internal cycle dwarfs the much

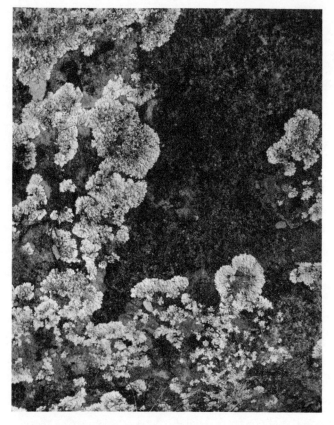

Living at the border of Gaia. These lichens are more chemically connected to deep waters around Antarctica than to minerals a fraction of an inch beneath the rock's surface.

smaller exchange with the environment. Thus even without knowing about our skins, we could identify the presence of our bodies as entities. The high cycling ratio of fluid—the discontinuity in rates between the internal cycle and the exchange with the environment—would give them away.

Through an upper border and exchange of gases, the lichen is linked to the atmosphere and thus to all soils, all waters, and all other life forms, and on relatively short time scales—a year for the air, a few thousand years for the ocean and soils, at most. But for the rock below the lichen, change is snagged on the permanence of crystalline minerals.

Moving from the lichen, which partakes of a rapidly circulating biosphere, to the rock, one passes from the gaian system of several thousand years across a discontinuity in time scales to a system whose scale of change is on the order of millions of years. Peeling away the rock's recently weathered surface reveals an ancient entity. Down a fraction of an inch within that bedrock, the minerals could have been the same for unimaginable ages. Eventually, ancient minerals and Gaia will meet; uplift or downward erosion brings them closer by about an inch every thousand years. At the lichen, the edge of sluggish bedrock coincides with the edge of a seething Gaia.

The borders of Gaia are neither as functionally discrete nor as crisp in space as those of organisms, endowed as they are with skins, membranes, or bark. But lack of these characteristics does not mean that Gaia's borders are difficult to define. At the outer border to space, the air fades into vacuum. The molecular density thins as altitude increases, and this occurs fairly uniformly along all radii away from the gravitational center. Little material is exchanged across this diffuse border, except for some hydrogen atoms and interplanetary space probes that leave, and some meteoroids and space dust that enter. The flux of energy is another matter, of course. A massive amount of solar energy enters; a similarly massive amount of infrared exits. The influx and outflux of energy both follow fairly simple patterns: parallel rays from the sun and spherically spreading infrared from Earth.

In contrast to the fuzzy fadeout of Gaia's upper border, the lower border looks crisp where the bottom of the lichen and the top of the bedrock meet. But consider what happens when ground waters enter cracks and pores of deep bedrock: Bacteria abound. And recent discoveries that have galvanized the life sciences have shown that bacteria can live miles below the surface, even in the pores of rock. Should the border of Gaia be extended to these deepest levels of life? Should Gaia's lower border be regarded as a fractal system of tentacles, an interdigitation with rock that reaches miles deep? Such a complicated border would follow along the colonized ground waters but then resurface to include the juxtaposition of lichen and solid bedrock.

Another tricky ambiguity about Gaia's lower border involves the type of rock that undergirds life. In many places, such as Indiana and Florida, the bedrock consists of calcium carbonate. This often whitish rock formed on continental shelves and under bygone shallow seas mostly from the burial of microscopic shells and anchored massifs of organisms such as coral. Deposited perhaps hundreds of millions of years ago, the shells have been compressed into rock. These calcium carbonates, which cover ten percent of the continental crust, were generated by life, so perhaps we should borrow a term used for biologically generated gases and designate the carbonates as biogenic rocks. Without life, on the other hand, physical processes alone would remove calcium from the ocean as calcium carbonate. Life precipitates the mineral on this planet before the pure physics comes into play.

A rock, biogenic or otherwise, is indubitably a rock. Once formed, rocks are usually long sequestered out of reach of the closely coupled, rapid-cycle system of the interconnected soils, air, waters, and organisms. Carbonates that date from the earliest days of life still exist; we study them in hopes of gleaning insight into Earth's ancestral chemistry. That's a long time to be out of circulation. Accordingly, carbonates and all other rocks should be regarded as outside the main system of Gaia. Outside as well are the remnants of organic matter called kerogen, locked in rocks such as shales.

I can find the lichen that inspired this train of thought by hiking up a nearby arroyo. This lichen's base is a particularly unambiguous lower border for Gaia because the question of how to interpret a biogenic origin for the rock never comes up: The orange-tan bedrock is pedigree igneous, born from great explosions of lava twenty million years ago. There is little if anything biogenic about a cliff of Bloodgood Canyon rhyolite or Bearwallow Mountain basalt. Such igneous bedrock issued forth not from the dynamics of biogeochemical cycles within Gaia, but from a force churning well below Gaia's lower border, however it is demarcated. The ancient Romans called that force a god — Vulcan. We call it volcanism and plate tectonics.

Pitted against Helios in an arm-wrestling contest, Vulcan would not stand a chance. The god of volcanism would be flattened before you could say "May the best god win." The flow of Vulcan's energy into Gaia, compared to that of Helios, is a snowflake to a snowstorm, a whisper to a heavy metal concert. In actual numbers, comparing the average of 60 milliwatts per square meter that diffuses up from the inner Earth to the 240 watts per square meter absorbed from the sun yields a ratio of 1 to 4000.

To an energy-hungry plant poised for photosynthesis, Vulcan's energy would be judged deficient not only in quantity but also in quality. Helios's high-frequency electromagnetic rays can kick up electrons from chlorophyll molecules and into life's complex photosynthetic pathways. Vulcan, on the other hand, emits only heat, and a low-grade heat at that. Except for rare spots where the temperatures from far below bubble up in hot water right at the surface, such as at geothermal springs, Vulcan's average surface energy would be negligible waste heat to a power plant engineer. Mere molecular vibration, diffusing up from Earth's interior, holds little promise for powering even atmosphere or ocean dynamics, let alone the gaian cycles of life.

Nevertheless, the inner planetary heat wields awesome power. Though it cannot kick off electrons in photosynthesis, it did bury Pompeii in ash flows and blow the top off Mount St. Helens. Vulcan can crack open oceans, split continents, bury ocean bottoms, and lift mountains. It slammed India into Asia and bulldozed up the Himalayas. The times scales are, of course, those of rocks: an inch or so a year for long strings of millions of years. As rays of light swirl the ocean, so heat rising from Vulcan can swirl the seemingly solid Earth. The power of Vulcan forged continents from what to us would have been unrecognizable configurations 400 million years ago. Consider, too, that whereas wind and water within Gaia, driven by Helios, can make molehills out of mountains, Vulcan can do the opposite. Vulcan's energy may be slight compared to the dazzling gift of Helios, but only the perseverance of Vulcan can make mountains out of molehills.

How can Vulcan do all this with so little energy? The parable of the tortoise and the hare comes to mind. Before reaching the surface, where it can enter the gaian system and radiate to space, inner heat from deeply dispersed radioactive minerals must diffuse upward through the mantle and into and through the crust. Thick continents, as thermal insulators, retard the flow and thus set up lateral differentials of temperature. Just as solar heat lifts air from the intertropical convergence zone, so deep heat from below lifts and circulates rock, which over long time scales moves like silly putty.

OK, Vulcan can flex some muscle and shift some rock. Does this do anything other than present an endlessly fascinating jigsaw puzzle for geologists? Does it do anything for Gaia? As a pre-existing boundary at the birth of Gaia, did Vulcan ever act on the entire living system in a manner that created a force of inward influence on the parts within Gaia? To answer this, we must examine the heat and motions of the deep earth for the exchanges with Gaia that are controlled by Vulcan.

Unlike Helios, which sends only energy into Gaia, the weak energies proceeding from Vulcan are not important for their energy per se. But Vulcan also affects Gaia by way of the material exchanges induced across Gaia's lower border. The yellow lichen on the bluff, for example, plucks several key nutrients from the rock's mineral storehouse.

The release of minerals from rocks by the various components of Gaia is called weathering. It comes in two flavors. One, physical weathering, involves the milling of rock from towering bluffs into microscopic grains. The other, chemical weathering, is less obvious but extraordinarily crucial. In chemical weathering, minerals are liberated from the rock and move as ions in water. Agents of chemical weathering include the acidic metabolites excreted by the lichen and the soils that retain rainwater like a sponge.

This supply of materials from Vulcan would be inadequate for today's gaian cycles of carbon or nitrogen, because, as we have already seen, these elements have global cycling ratios on the order of several hundred or more. But the elements that Vulcan brings to the surface in abundance have much lower cycling ratios. Calcium, for example. If someone daily gives you ten times the bananas you can eat in your perfectly balanced diet, you will discard most of them. The same seems to be true for the life forms evolved within Gaia. With ample supply from Vulcan, the tight cycling of certain elements is simply not a necessity for today's biota.

It would not have been enough for Vulcan merely to have provided an initial feedstock of rock at Earth's and life's beginnings. Fresh rock needs to be renewed, the mountains rejuvenated, the sediments out on the continental shelves hoisted high and dry and overturned. Vulcan's heaving exposes new surfaces for the small but crucial fluxes of minerals essential to the global biota. The continents act as wide open doors for these minerals' passage into Gaia.

Gifts from Vulcan to Gaia are present in the oceans, too. There, however, the sites are far more localized, more like cracks in closed doors. One crack is the 40,000-mile network of underwater mountains that snakes down the middle of the Atlantic, around Antarctica, and up the eastern Pacific, with numerous side branches such as that extending into the Indian Ocean. This ridge system is where new ocean floor, or crust, is born. It is where heat, fumes, and magma surface, as though from extrusion rollers, spreading in both directions, as the ocean floor parts.

Spots along the mid-ocean ridge feature the deep sea vents, from white to black smokers, where waters percolate up at temperatures that would be above boiling were it not for the extreme pressures. Expeditions to these vents with high-tech submersibles have revealed communities of exotic creatures: mouthless red tube worms, spindly crabs, and colonies of bacteria upon the rocky rubble and metallic sulfate chimneys. Although the total biomass of these creatures is low, the discovery of these wondrous communities has given us a wealth of information — and posed puzzling questions about chemical food webs, evolution, and the origin of life.

Less dramatic from the perspective of ecology but of utmost importance for Gaia is the circulation, through the entire ridge system, of water, which is driven into convection cycles through porous rock by the tremendous heat. The volume of this circulation is notoriously difficult to ascertain, but it appears to be enough to pass the entire ocean's equivalent of water through the ridges every ten million years or so. Dissolved ions react with the rock minerals as the water wends its way through the rock, entering cool and emerging hot. Magnesium is stripped, calcium added. Potassium is enriched, sulfate depleted.

What these fluxes mean for the global budgets, for ocean chemistry, and thus for life is not yet clear. For some elements, such as phosphorus, the ridge flux appears to be a minor contributor to ocean chemistry, removing far less than the burial fluxes from organisms and benthic precipitation. On the other hand, the exchange of calcium for magnesium, which would vary over time with the long-term (100- to 200-million-year) cycles of tectonics, seems to have caused well-established oscillations in the crystalline compositions of fossil carbonate shells—a direct effect of Vulcan on biological creations. The full importance of the ridge system as a two-way door between Gaia and Vulcan is still to be evaluated.

The sideways spread of the ocean floor at the ridges comes about not so much from a push of the upwelling material as from the pull at the other end of the plate-tectonic conveyor belt: the downward subducting slabs. Slabs are drawn down the deepest trenches, off Indonesia and Alaska and along the western coast of South America, where the sinking ocean floor forms a pocket with the continental material that remains afloat on the mantle. Pulled toward Earth's deep mantle by being relatively old, and thus cool and dense, the subduction zones would seem to represent the final removal of materials from Gaia. Over the tens to hundreds of millions of years it takes for deep ocean floors to slide from ridges into subduction zones, they accumulate red mineral dust from the continents and the white shells of carbonates and silicas that rain down as detritus from organisms at the surface. By the time these slabs reach a trench, a kilometer-thick layer of material from the gaian system may be traveling the tectonic conveyor belt to a destiny deep within Vulcan.

Yet the subduction zones can give as well as take. A geographical correspondence between subduction zones and volcanoes, either con-

Vulcan releasing gases. Anak Krakatoa—"Child of Krakatoa"—erupts in September 1979. Photograph by Michael Rampino.

tinental or marine, implicates the subduction zones as generators of gases purged upward through many miles of deep crackways in rock and into the atmosphere. Carbon dioxide, for example, can be volatilized and thus mobilized from its mineral prison in the carbonate sediments under high temperatures and pressures. Hence the soda springs, which release carbon dioxide, may be found at continental spots above the deep cauldrons fed by the subduction zones.

How efficient is the thermal cracking of chemistries? No one really knows. Lee Kump, a geochemist at Penn State University and a founder of the Geophysiological Society, has described the recycling via the churning within the sediment wedges at the subduction zones in terms of the ruminant stomach of a cow. These regions chew the cud of the ocean's crust and then regurgitate some of the materials—a portion processed into gaseous forms accessible to life—back into the gaian system. If this subsurface ruminant activity proves crucial to the maintenance of life, then because it belongs to Vulcan and not to Gaia per se, we have yet another superb example of an essential metabolic function that Gaia gets for free.

There is one more example of the relationship between Gaia and Vulcan. Note that the variety of Helios's gifts to Earth is rather limited. Parallel rays of light fall on a sphere; the resulting gradient drives the cycles of air and ocean and creates the grand climatic zones from tropics to poles. But the geometry of this gift is utterly simple, computable with a few equations. Vulcan, on the other hand, proffers a mind-boggling variety of gifts at the lower border of Gaia. Vulcan gives Gaia geology and geography. It determines the extent and placement of the two greatest habitats—continents and oceans. It is thanks to Vulcan that two places at the same latitude can be as different in climate as the dry mountains of New Mexico and the steamy lowlands of Mississippi. The heat of Vulcan that forces the various types of material heaves across the ventral surface of Gaia also ultimately creates the habitats for life, including the sandpaper bluff for the yellow lichen.

Though the surface of the planet deviates only slightly from a perfect sphere, Gaia crucially depends on the details of this deviation: the Tibetan Plateau; the Nazca trench off the western coast of South America; snowcapped Mt. Kilamanjaro just south of the equator in Tanzania; Death Valley in California, the lowest spot in the Americas. Topography is so vital to the structure of life on the planet that some have proposed that it join the pantheon of the spheres. In addition to the atmosphere, hydrosphere, pedosphere (soil), and biota, we should hail the toposphere, supplied for free by Vulcan.

⌐ The Union of the Trinity

The lichen on the sandpaper bluff lives within Gaia but is connected to Helios and Vulcan. From Helios it gets the high-energy photons for photosynthesis, warmth to speed its chemical reactions, and energy that circulates the atmosphere and thus links it to all life. From Vulcan it gets the bluff—habitat and source of nutritious minerals—put in place by eruptions millions of years before, chipped and dissolved away by the erosional attributes of Gaia, which expose fresh surfaces. And similarly to the lichen microcosm, Gaia as a whole resides between two great forces—energy presences—one dorsal, the other ventral.

The integrated system of Gaia, Helios, and Vulcan is quite different from that of the parts of our bodies, and not just because of the absence or presence of evolution. Recall one of the directives for gaian inquiry: Attend to the cycles of causes. We first applied this directive to the loops of factors that control photosynthesis and respiration within Gaia, but is it applicable across Gaia's borders? No. Helios does not participate in a loop of mutual influence with Gaia; the influence consists, rather, of a one-way arrow, from Helios to Gaia. We will have to keep an open mind about possible influences from Gaia to Vulcan (in mantle chemistry, for example). But Vulcan's deep, emerging heat is a primordial fact of Earth's geology that is independent of any gaian influence.

Although Gaia cannot touch Helios and Vulcan in the sense of changing the magnitudes of the energy fluxes that emanate from them, it can influence what happens to those fluxes at its borders. The most obvious example of gaian border control involves the amount of solar energy that enters the gaian system. At the top of the atmosphere, this amount is determined by solar output, distance from sun to Earth, and Earth's sphericity. But about a third of the energy that enters the atmosphere is reflected out to space by ice, by deserts, by molecules in the atmosphere, and most of all by clouds. These particular photons entered gaian territory but were never taken into the gaian metabolic system. They bounced back into space.

This example suggests how we should think about loops of influence involving Gaia, Helios, and Vulcan. We should focus on loops within Gaia, specifically those that connect the parts of Gaia that might be considered "doorways" with all its other parts. The presence of clouds closes some doors to the entry of the sun's energy. Vegetation, which absorbs more energy than smoother, lighter, more reflective surfaces such as rock or ice, opens some doors to the entry of solar energy into the system.

The general concept of a doorway can also be applied to the flux

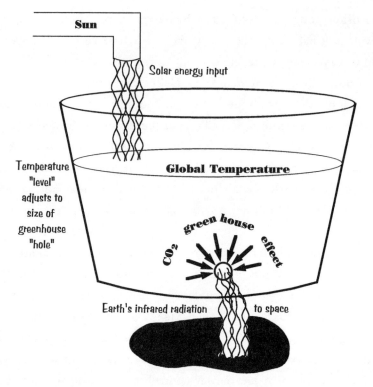

The leaky-bucket model of the greenhouse effect.

of infrared energy that departs Earth for space. The outgoing infrared flux, averaged over enough time, must equal the amount of solar energy that enters. Here the doorway—controlled by the greenhouse gases of the atmosphere—is analogous to a small hole in a bucket that is supplied with water at a constant rate by a faucet at the top: a leaky-bucket model of the greenhouse effect. Assume that the water entering the bucket and the hole are sized such that the water reaches a steady-state level within the bucket, a level at which the weight of the bucket's contents pushes water out at the bottom at exactly the rate at which new water is supplied by the faucet. If the hole is constricted, then the water level rises to a new steady-state position. The balance of fluxes is preserved: The higher water level (with its larger pressure) and the smaller hole create exactly the same outflow as the previous conditions.

When we expand this model to Earth, the entering water is the absorbed stream of energy from the sun, the height of the water is the temperature, and the exiting water is the infrared flux to space. The hole is the amount of greenhouse gases, where a smaller hole corresponds to a larger amount. The greenhouse gases are thus border elements of Gaia, and of course they exist in loops of influence with all the other parts of Gaia. The greenhouse gases (the hole) do not control the amount of the infrared flux (the exiting water), but they do determine what temperature (water level) is necessary to create a flux that balances the incoming solar flux (water from the faucet).

The concept of doorways under the control of Gaia is just as applicable at the ventral surface. Unlike the way clouds can reflect some of the solar flux, there is no way for Gaia to reflect any of the energy rising from Earth's interior; it all must eventually pass through Gaia to get to the infinite sink of black space. But the importance of Vulcan arises not from its energy per se but rather from what this energy does to the various fluxes of matter across the ventral border.

The heaves of Vulcan push rocks up to the surface, but how rapidly and completely they dissolve (rather than just crumble) is largely up to Gaia. Lichens dissolve rocks faster than air and rain do by themselves. This ability of life to intensify the rate of chemical weathering, and thus the entry of certain elements into the gaian system, is called the biotic enhancement of weathering, and it has been important in coupling biological evolution and atmospheric chemistry. Vulcan drives the delivery van of materials to Gaia's border, but the unloading crew steps forth from the biosphere itself—from life, but also from other factors such as the chemistry of the atmosphere and the amount of rain, both of which may be influenced by life.

The transfer of materials between water and rock at the ocean ridges also offers us an opportunity to examine controls exerted by Gaia at the ventral boundary. These complex chemical exchanges do not depend only on the minerals in the rocks upwelled into the ridges or on the heat that drives the water circulation through the ridges. They also depend on the chemistry of the water, and thus on the gaian system.

These fluxes are as yet so poorly understood that I hesitate to say more at this point. Nevertheless, by looking at water chemistry as a factor in controlling the fluxes of elements from Vulcan to Gaia (and from Gaia to Vulcan), we can ask questions about the borders of the gaian system that are not so different from the questions we asked about clouds as factors affecting the flux of solar energy.

Other opportunities for future investigations of Gaia's lower border will be found at those sites where Vulcan might effectively provide a ruminant stomach for Gaia—the descents of the ocean conveyors of plate tectonics, the subduction zones. At the very least, it is clear that what emerges from these hot, churning vats in the form of volcanic gases will be affected by what goes into them. Thus the relevant question is how the processes within Gaia affect the rates at which different types of buried materials enter the subduction zones. One example involves the difference between the fate of calcium carbonates that get buried as coral on the continental shelves, and thus remain on the continental masses, and the fate of those carbonates whose entombment occurs during the deposition of shells of free-floating plankton upon deep ocean ridges by the conveyor belt of sea floor spreading. The latter scenario holds far better odds for rapid resurrection via carbon dioxide belches from the tectonic vats of subduction zones.

Another potentially crucial issue involves how the presence of water itself may influence Vulcan. For example, the amount of water in the pores of sediments that enter subduction zones affects the viscosity of the sinking slabs, which would affect the patterns of continents and oceans. Does Gaia, then, influence the actual amount of water present as oceans? Life has helped Earth keep wet by consuming the hydrogen produced in water-splitting volcanism, thus preventing the loss of that hydrogen to space. But what fraction of today's ocean has been preserved because of early life? We don't know. But if it can be shown that Gaia influences the continent-making processes of Vulcan, then there would indeed be a loop of mutual influence, not all the way to the origin of Vulcan's heat, but to the doorways through which Vulcan influences Gaia.

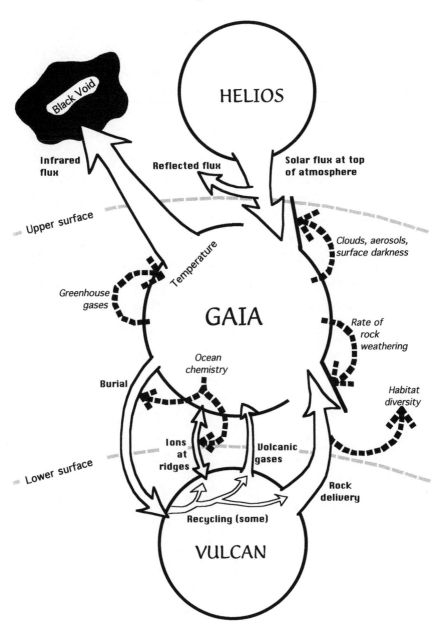

Interactions among Helios, Gaia, and Vulcan. Solid arrows show actual flows of matter or energy. Dotted arrows represent causal influences.

Much is given for free to Gaia across its dorsal and ventral borders by Helios and Vulcan. From Helios comes the energy to power photosynthesis (more on that later) and to drive the circulations of air and water that unite life into a system. The energy of Vulcan drives diverse fluxes of matter and presents a variety of geological and geographical habitats across Earth's surface. The unity of this trinity is not that of three equal partners involved in mutual support, like the three vertices of a structural triangle. Rather, Helios and Vulcan are what systems theorists call boundary conditions: conditions that act on Gaia. But there exists a deeper aspect to the unity, because, as we have seen, parts of Gaia act as doorways that control how much of Vulcan's material offerings and how much of Helios's energy gifts enter the gaian system. The important loops of causation are not among Gaia, Helios, and Vulcan but among the parts of Gaia, and some of these parts strongly affect fluxes from Helios and Vulcan. To those parts we shall now turn.

4
The Parts
of Gaia

I n late March of 1996, after disembarking from a bus at a bustling
square called Gloucester Green, I walked through a maze of ancient streets toward one of the colleges of Oxford University.
There Jim Lovelock and his wife Sandy were convening their second
gathering of specialists on Gaia. The first, on the same lovely grounds
two years earlier, had "the self-regulating earth" as its theme. This second meeting promised more of the podium to the biologists. Its theme:
"the evolution of the superorganism."

During the course of three days, most presenters dealt with details
about particular topics: fossil evidence for fire (and thus, by implica-
tion, atmospheric oxygen), effects of land life on rainfall, how patterns
emerge from collections ranging from neutrons to ants, the dynamics of

insulin regulation in human physiology. But Dan McShea, an evolutionary biologist at Duke University, took a chance. He stepped back from any current scientific fray and posed fundamental questions about wholes, parts, and Gaia. His working hypothesis was that as a number of smaller wholes combine to form a larger whole, they will shed some parts, as functions are taken over by the larger whole. He tested this idea by comparing the complexity of free-living cells such as amoebae with that of cells in metazoans such as we. The results were preliminary but positive.

At the end of the talk, I jumped up with an avalanche of comments. Topics dear to me had been addressed: first, the metapattern of wholes and parts (holarchies) and second, the parts of the unique whole called Gaia. McShea had pointed out that in trying to apply his or anyone else's hypothesis about the functions of parts within wholes to gain insight into the structure of Gaia, one would encounter an immediate problem. What *are* the parts of Gaia, anyway? Its organisms? Its ecosystems? Identifying two entities external to Gaia (Helios and Vulcan), as I did in the previous chapter, seems uncontroversial compared to delineating its internal parts.

The session chair responded to my outpouring by artfully muzzling me, using a time-honored technique. He suggested that the topic be picked up later in the day in a private discussion. Because a number of attendees showed interest in an impromptu group, we decided to gather that evening at a local pub.

↜ WHAT ARE THE PARTS OF GAIA?

A dozen colleagues had already begun the conversation when I squeezed into a seat at a long table of thick oak. Soon I was permitted to present a mini-lecture, hoping to stimulate and focus discussion about a number of issues I had not been able to raise that morning. The gist of my views follows. (Here I'll hold off citing comments from others until the end, but of course that is not how a scene in an English pub really goes.)

First, the parts of Gaia might be the largest-scale ecosystems, usually called biomes. This view offers the closest geometric analogy to our bodies. Just as the organs occupy unique locations within us, so the biomes spread across the planet in distinct biogeographical provinces: tropical rain forests, savannas, deserts, temperate grasslands, temperate forests, boreal forests, tundra. This dissection of Gaia into parts has immediate appeal because the viscera of Gaia would then be visual biological regions.

Shaping this concept into a substantial theory would first require that we delineate the biomes, each presumably unique. At first blush, nothing could appear simpler than distinguishing the tropical rain forests as a type, with their towering tree canopies, howling monkeys, strangling vines, and voracious ants. Or the grasslands, lush with slender, waist-high stems and herds of grazing animals. Or the deserts, unmistakable with their rough-skinned lizards, bare rocks, and spiny plants. Each type features a distinct look, the unique feel of an interconnected place, a subsystem of life on Earth.

But how many biomes are there? Fifteen? Seventeen? Thirty-one? Thirty-six? All these numbers have appeared in carefully reasoned scientific overviews, and all are used by various groups for analyzing the global biota (and that's usually focused on land!). Some divide temperature and rainfall into primary zones, which, crossed with types of vegetation, yield such sequences as boreal, temperate, and tropical forests or wet, moist, and dry tundra.

These neat categories belie the messiness of the details. For example, a savanna is traditionally an expanse of grass and other herbaceous plants, interrupted by patches of forest cover. What happens to a savanna when the fraction of these forest patches increases? The inventors of the land-cover classification for the International Geosphere–Biosphere Program specify a change in category from "savanna" to "woody savanna" when the forest cover becomes more than 30 percent, provided that the forest exceeds two meters in height. Then, when the two-meter-high woody cover surpasses 60 percent, the woody savanna turns into one of several possible "forests."

The problem of partitioning is very real for today's climate modelers. They must predict Earth's land cover in a warmer future with higher levels of carbon dioxide. Some reduce the divisions to a manageable number by treating temperature and rainfall as continuums imposed on vegetation predictions, thus limiting the discrete units to primary types of vegetation. One recent formulation, appropriately called the BIOME model, uses just seven categories: evergreen broadleaf, evergreen needleleaf, deciduous broadleaf, deciduous needleleaf (such as larches in Siberia), shrubland, and two types of grassland. Might then a small number of morphological life strategies of plants be taken as the parts of Gaia? Does Gaia particularly care whether a region hosts needleleafs or broadleafs—or even trees or grasses? Consider, for example, that nearly all of the story of life on Earth to date unfolded before grasses evolved.

The biomes are also poor markers of living functions at the planetary scale. It is not as though the rain forests produced oxygen, the tundra nitrogen, and the deserts phosphorus for each other and the rest of the biosphere. On the contrary, all biomes seem to perform all the functions, which then aggregate to a global scale. One or another biome may get out of balance with regard to particular cycles and thus depend on the whole to buffer its inadequacies, but each does contain the essential apparatus for cycling nutrients. Each biome, moreover, breathes the same atmosphere. Winds and waters can spread seeds and bacteria far and wide. The truest ecosystem in terms of a unique, bounded unit is thus Gaia itself.

For more fundamental biospheric components, we might turn to systematics. As the branch of biology that classifies organisms into nested taxonomic groups, systematics draws its facts from such fields as comparative anatomy, developmental biology, genetics, and paleontology. Groupings at the grandest taxonomic levels might also be the parts of Gaia.

One classification delineates five great "kingdoms": animals, plants, fungi, protoctists, and bacteria. Championed by Lynn Margulis for many years, the five-kingdom system uses cell type as a fundamental criterion of classification. Animals, plants, fungi, and protoctists are all eukaryotes (cells with nuclei, literally "good kernel"); bacteria are pro-karyotes (cells that lack nuclei and differ from eukaryotes in a host of other ways). In turn, animals, plants, and fungi—a familiar trinity even to lay naturalists—can be formally defined on the basis of their life cycles and additional differences in cell interiors. Protoctists remain somewhat as leftovers, consisting mostly of single-celled organisms (protists) such as algae and amoebae. But they also include some simple multicellular organisms that do not meet the criteria for membership in the other three eukaryote kingdoms, such as slime molds and seaweeds like kelp.

A mushrooming of new discoveries has generated the landscape of a classification even more encompassing than that of the kingdoms. Al-though one writer has suggested that we call the entities on this level "empires," thus preserving the political metaphor, they are most com-monly known by the more neutral term "domains." The domain of eu-karya (or, eucarya) subsumes the four eukaryote kingdoms. Indeed, they are all linked by a common evolutionary heritage. Thrilling evi-dence of this shared heritage has emerged, for example, in the abundant similarities between the cells of yeast and humans in the molecular ap-paratus of transcription from DNA to RNA to proteins.

What had once been the single kingdom of bacteria has now be-come two higher-order domains. The real bacteria are still bacteria in a domain of that name; they are sometimes also called eubacteria ("good-bacteria"). But a number of what had seemed to be oddball prokaryotes (many are literally balls)—anaerobic methane generators and others that thrive in extremes of heat, acidity, and salinity—are now known to be much more closely related to each other than to bacteria. Carl Woese, who pioneered this system of three domains, named these strange mi-crobes the archaea. Thus the three domains: eukarya, bacteria, and ar-chaea.

Should the domains be considered the parts of Gaia? Perhaps the domains, as types of genetic innovations, arose and proliferated because they were able to invent life strategies and thus occupy large-scale func-tional niches within the global system. Certainly from the view of the prime directive of gaian inquiry, the bacteria are essential: The biological world would collapse without them. The archaea had been regarded as curious and perhaps inconsequential recluses living where no one else would or could (deep sea vents, terrestrial hot springs, salt ponds). But now that the search for them is on, archaea are turning up everywhere—even in the ocean's depths that are comfortably cool. As for the eukarya, the world apparently worked fine for a very long time without them. For the first couple billion years of life history, only bacteria and archaea sustained the planetary system.

Overall, it is difficult to make a case that the three domains constitute useful categories for understanding the persistence of Gaia. So much biochemical diversity exists within each domain, and domains overlap in very significant ways. Photosynthesizers, for example, abound both in bacteria and in eukarya's kingdoms of plants and protoctists. The degradation of organic matter in soil provides another example. This vital task can be accomplished by certain bacteria, fungi, and even animals such as dung beetles. Although both domains and kingdoms are intriguingly deep structures of the biosphere's collective genome, functional overlaps seem to weaken any argument for viewing the groupings at either level as the parts of Gaia.

Are there other approaches? One, which we will consider further soon, looks at the four primary pools: life, soil, atmosphere, and ocean. Another possibility—somewhat more radical—is to consider the cycles themselves as the parts of Gaia. Pragmatically, this is how Earth system scientists tend to organize their knowledge. Books about the global biogeochemical cycles present chapters on the cycles of carbon, nitrogen, phosphorus, and so on. I implicitly used this principle of organization in Chapter 1, which featured carbon, and in Chapter 2 with nitrogen. The cycles as gaian parts will continue to be an underlying and virtually unavoidable theme in this book. If the cycles can be grasped as entities, then they can be assembled into a grand system for an understanding of Gaia. Advantages accompany this view. The cycles are more discrete than the genomic classifications can ever be—carbon is carbon, nitrogen nitrogen. The cycles of Gaia, moreover, are a manageable number, lim-

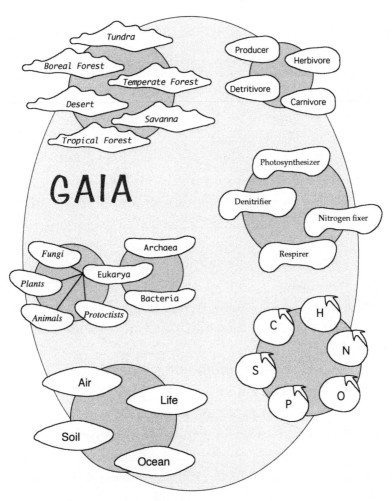

Viewpoints on the parts of Gaia. Biomes, trophic guilds, biochemical guilds, cycles, primary substances, genetic domains, and eukaryotic kingdoms. Perhaps the best view of all would be to conceive of the viewpoints themselves as parts that together make the whole.

ited to the essential elements (not the entire periodic table). And this system may quite possibly best reflect the way Gaia works.

Back at the pub, within the cantankerous group at the oak table, each of these different views attracted defenders and aroused critics. One fellow who was critical of designating the cycles themselves as the key parts argued that the cycles were more like flows and thus relationships among parts—whatever those might be. Another participant reminded everyone that the biomes should not be dismissed too hastily, because just posing the possibility of global functionality might lead to some new research questions. Witness the evidence from the fossil record that the tropics are the most productive latitudes for spawning new species.

The length of the table eventually exerted its own inward influence on the gathering. We split into two groups around the two ends, each group playing its own round of hacky-sack with the ideas. Gradually, and even reluctantly, my side moved toward a conclusion: a truth of no one truth. Instead, we decided, the truth resides in the very fact of the multiple viewpoints.

This view of multiple views was argued most succinctly by Richard Betts, a young British climatologist. He used the human body as an example. What are its parts? To a physiologist they are the organ systems: circulatory, respiratory, digestive, nervous, and so on. But to someone else, a carpenter or a dancer, the parts reflect the body's movement in space, and that implies head, torso, arms, and legs. There are functional truths in both views. Remember, Betts said, the viewpoints are exactly that: points of view that depend on what is being looked at and why. A neighbor at the table further posited that ultimate truth emerges from combining views, like transparent overlays, each atop the next. In fact, one might consider these very overlays the parts of the whole.

From the other side came a thumping of fist upon oak. Stephan Harding of Schumacher College in England had called for order. Harding wanted both sides to summarize their tentative conclusions before beginning the next round of creative bickering. Our side hesitated. Finally, I said that we decided to admit the validity of multiple viewpoints. A sigh of relief and then a cheer rose from the other group. With similar resistance and hesitation, they had reached the same conclusion. But Dan McShea, whose talk that morning had instigated this rowdy evening gathering, most resolutely held out for the possibility of a single best view. Perhaps he will yet turn out to be right. Otherwise, consensus was strong at that moment. Though not all views will prove equally effective, at least several will undoubtedly be potent modes of thinking about the parts within the holarchy of Gaia. Their power will prove out in the questions that flow from them. Beer glasses clanged together and banged down on the ancient table.

A few day later, the issue surfaced again in a public forum. Jim Lovelock rose to express his viewpoint. He suggested that the parts of Gaia are the major groupings of organisms that perform chemical functions in the global workings. In one of his mathematical models for Gaia through time, for example, he has three types of microorganisms: photosynthesizers, anaerobic methane generators, and oxygen-using consumers. In another model, devised with Lee Kump of Penn State University, algae of different species that produce dimethyl sulfide form a collective that affects Earth's chemistry and climate in Ice Age dynamics.

Lovelock's comments, though brief at the time, have ramified in my mind. Look again at three of the eukaryotic kingdoms: plants, animals, and fungi. They closely correspond to a trio of classic ecological functions: producers, consumers, and decomposers. Producers can be taken as synonymous with photosynthesizers, which transform solar energy and simple inorganic compounds in the environment into organic matter. Consumers use this organic stuff as a source of both matter and energy, and they create waste that is partly inorganic (carbon dioxide, for example) and partly organic (feces). Their organic waste, however, is generally not directly available to plants, which require nutrients in the form of dissolved inorganic ions and gases. That is where the decomposers step in. They complete the loop with a final transformation of organic wastes from consumers (and wastes from producers, such as dropped leaves) into those valuable inorganic ions.

This ecological trio is an abstract simplification, to be sure. But it serves as an initial map of the cycles of ecology, a baseline whose usefulness is apparent in how handily it can be elaborated. For instance, consumers are typically further disaggregated into those that feed directly on producers and those that stalk other consumers for a choice meal. In its simplest binary, this division yields the categories of herbivores and carnivores. Thus a cycle of three is partitioned further into additional levels (so-called trophic levels) of four or more—a food network or web.

In the parlance of ecology, more finely drawn divisions come down to guilds. Thus we have the guild of pollinators, which includes bees, moths, and hummingbirds. Other guilds are the bud-nippers, stemborers, root-chewers. For these one can almost imagine emblematic banners, such as those styled for the medieval guilds of pewterers, goldsmiths, and slaters. Like the workers' guilds in medieval economy, ecological levels and guilds are often considered the functional components. Note that neither species nor even higher taxa are the functional parts. Rather, the guild of fungivores unites subsets of the beetle and rodent families. Decomposers, a huge and general category, embraces fungi, many types of bacteria, and even some animals such as worms. Bacteria, fungi, and animals—three kingdoms and two domains—can thus be linked by the concept of trophic levels. Clustering by trophic levels and guilds, defined not so much by what they are as by what they do, may offer more insight into the parts of Gaia than the genetic groupings of kingdoms or domains.

Are we getting somewhere? The concept of guilds, as attractive as it is, presents its own difficulties. A squirrel is both fungivore and herbivore. A fish may graze on green plankton in its youth but hunt other fish as an adult. Such life-cycle switches and overlapping in classification encumber this approach. And note too that the strategy of using guilds or trophic levels for partitioning Gaia focuses primarily on the great restaurant of nature. Its emphasis is on who likes salads, who chomps live game, and who prefers offal. Nonetheless, this approach might guide our way to a division of the biota into various "organs" of Gaia. We might search for biochemical guilds. These would be ways of attending to the cycles of matter.

The seasonal oscillation recorded at Mauna Loa reflects a binary cycle between the dominance of photosynthesizers and that of respirers, which might be called two biochemical guilds in the cycle of carbon. Thus the concept of biochemical guilds in some cases cuts right across the divisions of kingdoms and domains. Respirers, for instance, include many types of bacteria, marine foraminifera, bread mold, and eagles. As with carbon, so with nitrogen. Its biochemical guilds would include nitrogen fixers such as the bacteria in soybean nodules and denitrifiers in low-oxygen sea water.

The functional parts of Gaia may therefore best be seen not by jetting across vast distances of latitude and longitude or by teasing out genetic affinities, but by standing in any one biome and simply looking around. Scan the plants and insects; pick up some dirt. All the biochemical guilds as the functional parts of biomes lie within reach of eye and hand. Certainly, the number of guilds can proliferate. For the carbon cycle, one would add the microbial producers and consumers of methane: the methanogens and the methanotrophs. The nitrogen cycle, too, is more complex than a binary system. Nitrogen fixers and denitrifiers would be teamed with the guilds of nitrifiers, ammonium assimilators, nitrate reducers, and ammoniaficaters. One can perform the same analysis for sulfur, phosphorus, and all nutrient elements essential for life.

Ultimately, one would have a list of the guilds for all biochemical transformations mediated by life. Would this exhaustive list embrace all the parts of Gaia? Could it be simplified by lumping? If a guild from one cycle always and only co-occurred with guilds from the other cycles, then lumping would be appropriate. Alas, the opposite is more likely. The combinations multiply the complexity. Some photosynthesizers fix nitrogen; some do not, instead directly assimilating ammonium and/or nitrate. Most respirers do not fix nitrogen, but some do. Sulfur could be added to carbon and nitrogen in the juggling of combinations. *Thioploca* is a sulfide-oxidizing, nitrate-reducing respirer. The bacterium below it is a sulfate-reducing, organic-nitrogen-assimilating respirer. Combinations of guilds would thus form superguilds, with long names like Spanish nobility. As combinations, the superguilds would increase geometrically. To be sure, some slots in the mix would remain empty, their hypothetical members not existing in nature. Also, this multiplying would come about only if we required each organism to be assigned to a single superguild. If organisms were simply permitted multiple memberships in guilds, defined for each essential element, then the numbers would remain reasonable.

Primary Pools

The cycles and the biochemical guilds are, to my mind, two of the best ways of partitioning Gaia to prompt provocative questions. They will thus be woven into the themes of the following chapters, just as they have informed previous discussions. But there is one additional overlay, which I've waited to elaborate on until now, whose value as a mind tool rivals that of any other. This is a partitioning based on primary pools.

The search for primary substances is nearly as ancient as civilization itself. The Greeks had four: earth, air, fire, and water. The Chinese system had five: metal, wood, earth, water, and fire. Such substances, or so-called elements, exist in relatively pure forms and, according to the ancient theories, can be mixed to compose the sundry things of the world. It is possible to see such theories as ancestral to our current understandings, by which subatomic particles combine to form the suite of elements, and by which the atomic elements combine in turn, yielding the bounty of molecules and minerals.

The primary substances of Gaia, I maintain, can be even simpler than the parts of such alphabet-like systems. They are all, for example, visible from a beach: the wide, mysterious ocean, the sand and soil underfoot, the vines that cover the dunes behind, and the sweet air all around. These are more like the spatially distinct organs in our bodies than like the chemical elements that combine. They are close, however, to the ancient four "elements" of earth, air, water, and fire, especially considering how often fire is used as a metaphor for life. Indeed, the big four pools of Gaia—ocean, atmosphere, soil, and life—do mix, and this proves crucial, for mixing integrates all into a planetary system.

A disclaimer to start: I would not bet the life of any one of my intestinal bacteria on there being any ultimate, logical imperative for designating these four as *the* gaian pools. Maybe there are five, maybe six. Maybe more. Maybe thinking of systems in terms of a small number of biggest parts serves only as a helpful heuristic for our thoughts and investigations. Perhaps the ocean should be separated into coastal and open ocean waters or into water and sediments. And should life itself be treated as a single pool? If not, then some of the previous views present ample choices for primary divisions of life. So take the four gaian pools with a grain of salt, but nonetheless savor all four courses.

Let's begin with life. The Russian V. Vernadsky, a founder of biosphere science, praised life as "the most powerful of the geological forces." Life is the central component of the prime directive for gaian inquiry. Using a term from alchemy, life is the philosopher's stone. Modifying a phrase from ecology, life is the keystone pool.

How can life be considered a pool of a particular substance? That simplification may seem too ambitious. We all do speak the word, however, which demonstrates a collective nod to what all living forms share, an underlying unity; such understanding is irretrievably embedded in the early learning of the word by every child. This is one case in which the more we know, the more the root perceptions that gave rise to the word branch lushly. Behold what has been found by scientists who delve into the microscopic and molecular details of life.

As I write, perched on a rocky ledge above the river, a tiny bug — so dainty — is crawling across the page. It's about two millimeters in its largest dimension. I'm sure that even with the bifocals I am sorely beginning to need, I could hardly better resolve the two dots that must be its antennae. Where are its legs? And yet it walks. I try to coax it onto my finger to look for those minuscule appendages, but it freezes and slips down the page like a skier on a slope. That catatonic freeze suggests that it is a soil organism, one whose name, were I to see it printed, would be difficult to pronounce with assurance because in my entire life I would never have heard it spoken. So I'll name it. Good luck with your journey, *Cautious invisipod*.

So different and yet so similar! Its rigid, chitinous shell contrasts with my supple skin of keratin. Yet both tough border materials, chitin and keratin, were manufactured by cells—both bug and I are cellular beings. My cells and those of the bug are bounded by membranes of lipid molecules, and nearly all have nuclei with DNA (some exceptions are my red blood cells). All have, strewn throughout their interiors like raisins in ice cream, collections of kidney-shaped organelles called mitochondria, which create energy for the cells in the form of molecules of adenosine triphosphate (ATP) by way of the Krebs cycle. Thus a spectrum of similarities between bug and me ranges from our cells on down the holarchy to a vibrating netherworld of the universal molecules of life.

Mitochondria and nuclei are found exclusively in the domain of eukarya. More ubiquitous are the ribosomes, present in the cells of all three of life's domains. Ribosomes are tiny and numerous knobs of protein and structural RNA in all eukarya, all bacteria, and all archaea. Like railroad yards where cars are linked one by one into a long train, ribosomes are assembly sites where small molecules of amino acids are linked into giant, extruded chains of proteins. Differences between ribosomes (specifically, differences in their RNA) enable geneticists to group organisms into genetic affinities—clusters of evolutionary relationships that map the existence and contents of the domains themselves.

Harold Morowitz, author of *Energy Flow in Biology* and *Beginnings of Cellular Life*, has pondered these and other universals of life. He calls such grand facts biological generalizations. They include the cellular structure of life, the essential presence of water in life, the ribosomes, the key role of phosphorus in the currency of molecular energy, the universal lipid membrane, the catalysis of reactions by enzymes, and the genome of DNA.

One prominent generalization in Morowitz's list concerns a core set of six chemical elements in the bug, me, *Thioploca*, and all life we know. These elements form the bulk of all biomass and the structural hearts of biological molecules. They all can enter into covalent atomic bonds (bonds in which electrons are shared between atoms, rather than hogged by one atom of the bond). The six elements—carbon, hydrogen, nitrogen, oxygen, phosphorus, and sulfur—can be remembered by the acronym CHNOPS, pronounced to rhyme with the ancient Egyptian pyramid builder (thus CHee-NOPS). Their proportions are remarkably constant across the entire diversity of life.

The holarchy of stratified stability extends outward from the elements of CHNOPS to small molecules formed by combinations of those elements, and ultimately to the largest macromolecules. The universal small molecules include finite subsets of amino acids, ribonucleotides, deoxyribonucleotides, sugars, and fatty acids. Morowitz points out that 90 percent of nonaqueous cellular material consists of fewer than fifty such molecules and their combinations into the four main classes of macromolecules: proteins, nucleic acids (RNA and DNA), carbohydrates, and lipids.

Such generalizations add enormous weight to the assertion of a common ancestry from which all life stems. If evolution is variations on a theme, different choruses sung using the same scale of notes, then the totality of life can be considered a substance. Surely the similarities in composition among different permutations of life should guide our thinking about how organisms fit within the whole. As one of his biological generalizations Morowitz explicitly emphasizes the intimate link between individuals and a much larger level of the holarchy:

> In nature, individual organisms do not exist in isolation . . . life, as individual organisms, persists over long geological periods only because of integrated processes of the atmosphere, the hydrosphere, the lithosphere, and the biosphere. In that sense life is a property of planets rather than individual organisms. This thought has been expressed in James Lovelock's presentation of the Gaia hypothesis.

On land, when life leaks and is transformed into another type of primary part, it does not have far to go. Gravity may take it down as leaves tumble or entire trees topple, or as seeds disperse, settle, and then perhaps fail to germinate. Cast down, too, are the uneaten remains from animals' feasts (the feathers and split acorn shells) and the ingested but undigested lumps excreted to join the leaves and seeds at rest. To these must be added the roots, fungi, and bacteria that are already in place for the transit into material within the new pool, after their life force dwindles.

Vulcan as well as Gaia accounts for the genesis of the second primary pool, which blankets the continents as a nourishing placenta for life. From Vulcan's bedrock come broken bits. Physical weathering chips these bits smaller and smaller into gravel (which the poet Seamus Heaney calls "milt of earth") and then grinds them further into sand, silt, and clay. Passage along this gradient of sizes is aided by chemical weathering, which transforms the composition as well, all the more so as the bits approach the bacteria-size specks of clay.

Remnants of the formerly living also pass through a sequence of diminishing sizes and altered compositions. Devoured by crawling and burrowing scavengers, physically ground and chemically broken by worms, beetles, ants, and invisipods, the decomposing biomass enters a complex web of life that includes tiny worms called nematodes, whose populations per square meter often number in the millions, each of which can gobble thousands of bacteria per minute. Debris from fungi and bacteria are soon encrusted with inorganic coatings, forming "humic" materials. Acids are secreted by plant roots as well as by a variety of microscopic life forms. All together make a matrix of different-size aggregates—sticky, sponge-like, complex, nourishing.

The result is the primary pool soil. It comes in more colors than human skin, and also in flavors. (Experts can recognize soils by their taste.) Its various names—alfisols, aridsols, mollisols, oxisols, and many more—like those of biomes, help us organize the bewildering gradations of acidity, mineral contents, and amounts of organic matter.

Soil can be regarded as a substance of planetary physiology, despite so many differences across regions, because it passes the same test we used to confirm such status for life: its own parts are much the same everywhere. A key aspect of soil's fertility derives from life itself. The four main classes of macromolecules, the less than fifty small molecular units, the six elements of CHNOPS—upon the death of organisms such biological generalizations create a generalized soil. Soil everywhere receives a flux of fairly similar dead life, which is then mixed, digested, and aggregated with particles of minerals into a matrix.

From my perch above the river I see a young juniper tree, already gnarled from its struggles on the cliff for structural purchase and water. The trunk and fallen scale-leaves around its base catch dead grass stalks, sand, and silt—the building up of soil has begun. It will not progress far on this rough outcrop, doomed by the site's inability ever to hold enough matrix to buffer against sterilization by drying or against erosion by a ripping rain. Nevertheless, for now some forbs have found satisfactory protection in the circle of fertility around the tree.

Soil is where crucial links churn in the cycles of biochemicals. To know anything about the cycles, one must know soil and its inhabitants. Because the decomposed products of life make soil so much of what it is, without life there is no soil. Sediments, yes, but no soil. Without life, the forces of weathering on this cliff would create particles of different sizes, but these would be whisked or whooshed away. Without life and thus soil, every sunny day would be a drought, every rain a flash flood. Without life and thus soil, particles would quickly wash or blow away, to aggregate in valleys in deep sediment graveyards or tumble into the sea. Much of the continents would be as bare as today's fresh volcanic fields. This mutual influence between soil and life has prompted the Dutch soil scientist N. van Breemen to call soil a "biotic construct favoring net primary productivity."

Stirred into uniformity by global swirls, the air is ostentatiously a primary pool. But whether this third primary substance of Gaia is also a biotic construct favoring the productivity of life is a more controversial question. Consider now just the first part of the question: Is the atmosphere a biotic construct? Invoking our prime directive we must ask what the composition of Earth's atmosphere would be without life.

No one knows in every detail. The loops of feedbacks are too complex. But two decades ago, James Lovelock delineated the major differences between the atmosphere of a nonliving and that of a living Earth.

Virtually gone from a nonliving Earth, for a start, would be the so-called biogenic trace gases. Somewhat exotic by-products of certain groupings of living things, the biogenic trace gases include methane (CH_4) and nitrous oxide (N_2O). These are the two most important greenhouse gases after water vapor and carbon dioxide. Another crucial biogenic gas is dimethyl sulfide (CH_3SCH_3), which is apparently the major source in marine air of invisible acid droplets that seed cloud formation.

Without life, furthermore, the second most abundant gas, oxygen (O_2), would be as scarce as a winning lottery ticket. In photosynthesis, water is split into oxygen and hydrogen. The hydrogen is joined to carbon dioxide to form biomass (schematically, the simple sugar $C_6H_{12}O_6$). The oxygen is freed as a gas. Respiration of the biomass reverses the process and thus uses up the oxygen that was just made. Question: how is net oxygen generated? (Moreover, it must be generated continuously. Without a constant source, the oxygen would slowly but inexorably disappear into oxidizing reactions with rock minerals and volcanic gases.)

Answer: the generator of net oxygen is photosynthesized biomass that is never respired. This happens when biomass is buried and compressed with minerals, eventually darkening rocks such as shale. But compared to the gush of photosynthesis, the flux entombed is a trickle indeed, about 1 part in 2000 of the organic carbon made in photosynthesis. Nevertheless, each atom of buried carbon supports the retention, in the air, of a molecule of oxygen gas from life.

What would nitrogen (N_2, 78 percent of air) be without life? This is difficult to answer because, as we have seen, life powers both sides of the cycle through two of the biochemical guilds of nitrogen: denitrifiers and nitrogen fixers. But what would happen in a world with neither guild of bacteria? Well, there is a process that can fix nitrogen in the absence of any life. Enter the thunderbolts of Zeus.

How much nitrogen is fixed by lightning is difficult to assess. Although research tools have progressed considerably since the days of

Ben Franklin and his kite string, capturing the violent chemistry of a thunderstorm is still an iffy and even dangerous pursuit. Lightning rips the triple bond in the nitrogen molecule. It's like going by bicycle up a small hill to cruise down a really big hill. The big downhill is the formation of the final nitrate. In brief, the storm's electrical energy prompts molecules of oxygen and nitrogen first to split and then to fuse into two molecules of NO, which then further react with oxygen to form nitrate. Rain washes the nitrate out of the atmosphere.

Thus if all life were removed from the planet, denitrification and biological fixation would cease, but fixation by lightning would continue. Remember, though, that the circulation of ocean water through the hot ocean ridges would probably denitrify the nitrate back into nitrogen. The atmosphere's nitrogen in an abiotic world *could* be close to today's amount—with emphasis on the *could*, because so much of the world's chemistry would be altered beyond recognition without life, and nitrogen has one of the most complex chemistries of all elements. Note, however, that the fixation by lightning and the thermal denitrification would be ten to twenty times less vigorous than the analogous fluxes driven today by biology.

Carbon dioxide concentrations, now nearing 365 ppm and crucial because of greenhouse properties, would also be very different without life. Unlike N_2 and O_2, rather than nearly vanishing, carbon dioxide would reach an enormous level and engender a hothouse world. The topic of life's impact on CO_2 by the biotic enhancement of rock weathering will be revisited. For now, it is sufficient to recognize that the atmosphere is remarkably much a biological construct: Without life there would be virtually no oxygen, slower nitrogen dynamics, almost nothing of the trace gases that are primarily biogenic, and a greatly elevated level of carbon dioxide.

The fourth primary pool of Gaia is close to an element in both ancient Chinese and Greek philosophies. I like to identify it as ocean, and to let some of the other, extended presences of water be parts of atmosphere, soil, or life, as appropriate. Various options for drawing boundaries might be allowable, just as long as some of the "extras" described below are not neglected.

One reason for identifying the fourth primary pool with the ocean itself is simply that so much of all water is there. Of every million molecules of planetary water, the ocean contains about 972,000. Glaciers and the giant ice caps of Greenland and Antarctica sequester another 21,000 frozen molecules. The share in all ground waters, down to two miles or so, is about 6,800. Capping off the million, and laughable in amounts but not in importance for Gaia, are 130 in lakes, 60 in soil moisture, 9 in the atmosphere as water vapor and cloud droplets, a mere 1 in all the world's rivers, and, finally, just about half a molecule in the totality of life.

Another reason for preferring the name ocean is the need to consider a suite of extras within the water itself. Evaporating a bit of it makes the invisible ions visible as white crystals of salt, about 3.5 percent of its mass. The ocean's top four dissolved ions—in order of abundance: chloride, sodium, sulfate, and magnesium—constitute 97 percent of its total salt content. All elements, however, are present in their particular ionic species. Our bodies contain a fluid that is called plasma when it is in the blood system and extracellular fluid when it migrates to bathe all cells. In it chloride and sodium are likewise the top two ions. Though their order of abundance is reversed and their concentrations are only about a fifth of what they are in the ocean, they illustrate the crystal of truth in the statement that we carry the ocean around within us.

Mixed among the ocean's ions and water molecules are particles that span a huge range of sizes. At the fine end are life's small molecules such as amino acids, macromolecules such as lipids, and tiny bits of cell parts (if they pass through a half-micron filter, they are collectively called "dissolved" organic matter). Particles caught by the filter include larger parts of cells, whole cells themselves, and fecal pellets falling from tiny invertebrates. Also entrapped are the visible, downward-drifting, amorphous accumulations of everything and anything, aptly named "marine snow." Particles from land, too, join the chemical mélange via rivers. And winds transport fine dusts everywhere, even in measurable amounts to the middles of the oceans, where they slowly settle to form portions of the deepest sediments. With all these assorted extras, the ocean is like soil: a complex matrix of offerings from both the abiotic and the biotic realms.

As with soil, life impresses certain chemical patterns in the ocean. One celebrated example involves the concentrations of the nutrient ions nitrate and phosphate. The average concentrations of nitrogen and phosphorus masses for the world ocean are in a ratio of nearly 7 to 1: roundly, 400 milligrams of nitrogen in nitrate to 60 milligrams of phosphorus in phosphate, per cubic meter of sea water. Moreover, concentrations of these nutrients at any given point in the ocean are consistently close to this ratio. Thus in a cubic meter of water from the deep Atlantic that has the relatively low 30 milligrams of phosphorus in phosphate, we will find 200 milligrams of nitrogen in nitrate. In the deep Pacific the waters are more "aged," because they have been receiving a flux of nutrients recycled from the falling organic detritus for a longer time since they were last "reset" to zero by photosynthesizers at the surface somewhere. They typically hold fifty percent more phosphate than the global average. As sure as the rising sun, those waters also contain fifty percent more than the average nitrate.

Why the perfect tracking of nitrate with phosphate? A hint of the answer to this mystery has already been given. It's in the falling detritus and other productions of life, such as the dissolved organic matter. Measurements of a large number of species of algae and zooplankton have shown that nitrogen and phosphorus occur in the same 7-to-1 mass ratio in their living bodies. But how does this ratio become impressed on the ocean's chemical signature in the presence of large fluxes such as nitrogen fixation and denitrification? That question will be answered in detail in a later chapter. For now, we simply acknowledge that the generalized composition of life thus contributes to generalized patterns in ocean chemistry.

⁓ Geometries for Global Metabolism

Four primary pools, or substances, have been set forth and briefly surveyed, particularly for the impact of one, life, on the other three. But that is not enough. Life, soil, air, and ocean—the four form a system, and surely it is the interconnected dynamics of the whole system that we wish to comprehend.

I submit that these four are somewhat more than a useful, low-number heuristic. Rather, together they express something about the reality of Gaia, something that will serve as a framework for developing insights into the dynamics of the whole system. For instance, we could start with the four pools, begin overlaying one or more of the other possible views of the parts of Gaia, and in this way build toward answers to particular questions.

A potentially fruitful choice is to add the biogeochemical cycles to the primary pools. That alone is a daunting endeavor. But a long journey begins with a single step, or, in this, case, a relatively simple question. How are the crucial elements distributed among life, soil, air, and ocean? For now, let's focus on the C, N, P, and S of the six key elements (CHNOPS). Hydrogen and oxygen are simply too overwhelmingly abundant in ocean—linked as water molecules. The distributions of the remaining four present more subtle puzzles.

Sulfur, least of the four in terms of concentration in biomass, is still absolutely essential for life—for example, in some amino acids, the building blocks of proteins. Where is Gaia's sulfur? A whopping 99.98 percent is in the ocean waters. This fact is graphically displayed in the figure. Note that none of the other pools even comes close to showing up on sulfur's diagram (0.5 percent is the minimum for display). Sulfur is extremely soluble as sulfate (SO_4^{2-}), the ocean's third most abundant ion. Its most copious form of gaseous flux from the ocean, the biogenic trace gas dimethyl sulfide, fairly quickly becomes oxidized to sulfate, dissolved in water droplets of clouds, and rained back down. Similarly, sulfur from combustion pollution is rapidly stripped from the atmosphere as acid rain. The high solubility of sulfate keeps these

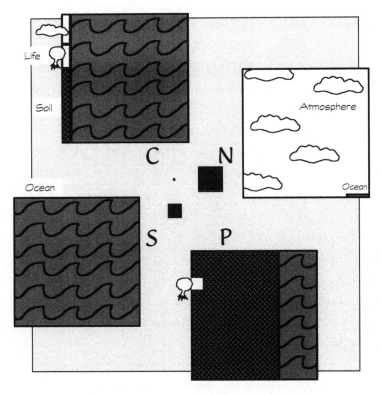

Distributions of elements among pools. Carbon (C), nitrogen (N), phosphorus (P), and sulfur (S) within the pools of atmosphere, life, ocean, and soil. In each case, only those pools that contain at least 0.5 percent of each element are shown. The black squares in the center show the total amount of each element, summed over all its pools, relative to the other elements.

circuits short and quick and thus severely limits the amount of sulfur in the air.

Next is an element even more crucial, if measured by its abundance in life's amino acids, and one whose pattern of distribution is as skewed as sulfur's. This is nitrogen. The atmosphere pretty much has it all (99.4 percent). The ocean's allotment claims a small corner of the diagram (just a bit below 0.6 percent), nearly all as dissolved nitrogen gas. You may think that does not leave much for the ammonium and nitrate ions of soil, the nutrient nitrate of ocean, or the organic nitrogen of life, and you're right. Not surprisingly, nitrogen is the bulkiest part of fertilizer applied to farmlands.

Turning to phosphorus we find that the soil pool not only registers above the half-percent level but even dominates. Sixty-eight percent of Gaia's stock of phosphate resides in the soil. The ocean contains a hefty share as well, at 31 percent. What about the atmosphere? Except for a presence in dust and perhaps one or two very rare gases, phosphorus lacks a place in the atmosphere.

With phosphorus, life finally makes an appearance on the diagram, at 1 percent. Phosphorus is universal in such cellular ingredients as the sugar phosphates of the DNA helixes and the molecules of energy exchange. As already noted, the concentrations of phosphorus and nitrogen in life are related by a mass ratio of only about 1 to 7. The reason for life's visibility in the inventory of phosphorus but not that of nitrogen derives from the relative paucity of phosphorus in Gaia. The little fish looks bigger in the smaller pond.

Finally we come to my favorite of the four, an element that has been the focus of much of my career. I have saved carbon for last not because it's been my bread and butter, or even because it forms the molecular spines and rims of all organic molecules. Rather, with carbon, each of the four pools accounts for more than the half-percent threshold. The ocean's carbon, mostly in the form of bicarbonate ions, takes first place. The allocation of the remaining seven percent can be remembered by means of simple ratios. First, about equal amounts of carbon reside organically in the biota and as CO_2 in the atmosphere. And soils contain about three times the amount of either the atmosphere or the biota (globally averaged, of course, and primarily in the top meter or less). I know these numbers better than the balance in my checkbook.

The inventories of the four elements lead to several noteworthy observations. First, the ocean, soil, and air are clearly of equal importance, despite their unequal appearances and internal masses, in the sense that each is the number-one pool for some element. The ocean contains most of both sulfur and carbon. The atmosphere takes precedence for nitrogen, the soil for phosphorus. Thus one can no more be a Gaian without understanding ocean, atmosphere, and soil than a Catholic without recognizing father, son, and holy spirit.

The second observation highlights the opposite end of the scales. Life comes in last for two elements: nitrogen and carbon (most accountants of the carbon budget peg the total carbon in life as somewhat less than that in the atmosphere). For phosphorus and sulfur, life occupies position number three. And it is mostly land life at that. Marine life accounts for less than one percent of global biomass.

Given its low rankings in our inventory of elements, should life be ignored? Certainly marine life could be disregarded, no? No. We have already seen enough of gaian dynamics to know that Earth without life would be a very different Earth. For starters, recall how the seasonal cycle of photosynthesis and respiration causes the oscillation in the atmosphere's carbon dioxide. And how the denitrifying bacteria created conditions that prompted the evolution of nitrogen-fixing bacteria, thus making possible the high productivity of today's land life. These facts are invisible in the static distributions of the elements. More generally, consider that the soil is largely a biological construct, that much of the air is a biological construct, and that significant features of marine chemistry are biological constructs. Beyond question, life, despite the fact that its repositories offer mere thimblefuls of storage, cannot be ignored. It even turns into the keystone when we start to evaluate the relations among the primary substances—in other words, the fluxes as well as the seemingly static pools.

Geometrically, four pools paired in all possible ways yield six relations. Each relation between any two will have several components: fluxes of matter, fluxes of energy, and influences. For the most part, the fluxes and influences take place at physical interfaces between pairs of pools. A crucial item to examine, therefore, is the interfacial surface areas.

The border between ocean and atmosphere totals nearly seventy percent of the surface of the Earth sphere, more than 35 times the area of the United States. Gases flow to and fro, like the sharing of breath between the lips of lovers, at this vast interface of water and sky. Nitrogen, oxygen, carbon dioxide, and others enter and leave in prodigious quantities via what is called air–sea gas exchange. Elements may escape and return in different forms, such as sulfur wafting up as dimethyl sulfide and falling back as dissolved sulfate in rain. Energy transfers across this surface are important, too. Winds pick up heat from the generally warmer sea, and the evaporation of water vapor moves energy as well as matter.

The second relation, that between soil and air, occurs across the remaining thirty percent of Earth (including the ice, for simplicity). This land surface is about 60 million square miles, an area over which the human population could be spread to an average density of about a hundred people per square mile. Between soil and air, gases flow in an exchange analogous to that between air and ocean. For example, the bulk of the net respiration in fall and winter that powers the upward pulse of the Mauna Loa oscillation is generated in the soil and then outgassed into the atmosphere. Evaporation, rain, and heat transfers make the soil, once again, similar to the ocean in many ways.

A third pairing connects soil and ocean. Except at the edges of continents, these two do not directly meet, but they are indirectly coupled through their consort with the air. Wind-blown dust travels to the sea. Water vapor from either land or sea can rain down over the other. In addition, sluggish ground waters and roiling rivers carry to the ocean dissolved ions, mineral particles, and dissolved and particulate organic matter. (I include the ocean's sediments as part of the ocean pool—an important part that Vulcan periodically turns back into land, renewing the cycle.)

The final three relations of the six all involve life: life and air, life and soil, life and ocean. The complexities of these relations have already been set forth in the examples used to demonstrate overall influences of life on the other three pools. More will come. But for now, I want to highlight one factor that imbues life with much of its power in the gaian system. Grasping this factor is the first step in understanding how such a tiny player as life can have such a big kick in the fluxes of the biochemicals. Ultimately, the exchanges between sea and air and between soil and air are functions of their interfacial surfaces. So too with exchanges involving life. And life has strategies for creating enormous surfaces.

Let us begin at the beginning of the terrestrial food web, with the biological devices for absorbing sunlight and converting photons into the chemical energy of biomass. Whether broad or needle-like, leaves are functionally surfaces for the absorption of light. Leaves also exchange gases with the air. Terrestrial ecologists quantify these green surfaces in a given forest or biome and thus calculate a leaf area index. This index expresses the number of ground area multiples that different types of vegetation exhibit as total leaf area.

In the BIOME model already mentioned, seven types of vegetation vary, in their average, two-sided leaf area indexes, from about 2 for shrubland to nearly 9 for evergreen broadleaf forests such as rain forests. Summing over the respective areas, including the twenty percent of land that has a leaf area index of zero (because it is ice or nearly barren desert), yields a leaf area that is three and a half times the world's total land area. That figure is easy to remember, for it is just slightly more than the surface area of Earth. Thus life's total leaf area is equal to 1 Earth area.

In the ocean, most photosynthesis is performed by a suite of algae and bacteria of various sizes. (We'll consider bacterial surfaces separately later.) The total carbon in the biomass of algae, compared to that of land leaves, is minuscule, so if algae clumped their limited mass into leaves, their surfaces would be negligible compared to those on land. Instead, the ocean floaters have chosen another strategy from the repertoire of geometric surfaces: to be small and numerous. Imagine dispersing all the cells of a leaf into the ocean where each cell would bob as a individual ball. The total exterior surface would be greatly magnified.

Using global average estimates for masses and sizes, I compute that the total surface of this green marine living substance is on the order of 3 to 8 Earth areas. These numbers make sense. Recall that marine photosynthesis is roughly equal in amount to that on land (again,

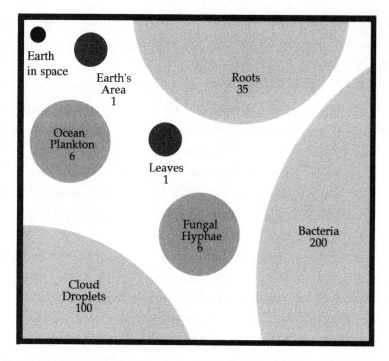

Surface areas. Circles show the sizes of areas relative to the surface of Earth—that is, in Earth areas. When estimated high and low values are cited in the text, the averages are shown here.

despite the relative paucity of total marine biomass). But the floaters are thinner and thus more transparent than leaves, making the amount of actual absorption per unit of surface less. Also, whereas some leaves capture direct sunlight, algae must deal with the blocking effects of intervening water. Thus one would expect that a given flux of photosynthesis would require more surface area when it relied on tiny ocean spheres rather than big terrestrial sheets or needles.

Back on land, in the soil toil organisms that excel in string-like geometries. Briefly mushrooming into surface prominence, fungi spend most of their lives as microscopic, colorless hyphae. Fungi are decomposers whose threads form spaghetti-like nets through the soil matrix. These surfaces secrete enzymes that digest remnants of life—even lignin and chitin, which are too tough for most other organisms to dismantle chemically. Digested and dissolved nutrients are then absorbed into the hyphae. The geometry of hyphae is extremely complex; for example, a functionally significant amount of hyphae exist symbiotically inside the roots of plants. Globally, soil fungi carry an internal pool of organic carbon about equal to that of marine plankton. Cross sections of hyphae are roughly the same as those of the bodies of ocean photosynthesizers. Disregarding for a moment the geometric difference between tubes and spheres, these similarities would lead one to estimate that both groups have about the same surface area. Indeed, my best calculation yields a global surface area of 2 to 9 Earth areas for these impressive underground foragers.

Also in the soil, and employing the same tubular strategy as fungi, is another category of surface area maximizer. Visible roots impress us with their lengths, but the actual surfaces for absorbing water and dissolved nutrients are microscopic extensions called root hairs, put forth by the outermost cells of the finest roots. Estimates from plant ecologists place the surface area, summed globally, of all root hairs in a range from 12 to 60 Earth areas. Although their diameters (ten microns, or about one-fifth that of the typical human hair) are about the same as those of fungal hyphae, root hairs beat hyphae as global generators of absorbing surface on the simple strength of their greater mass.

Finally, there are the bacteria, masters of the geometric tactic of being small and numerous. (I have no separate estimates for archaea, so we must crudely lump them with the bacteria for now.) How much bacterial mass is there? No one knows. One estimate from the carbon accountants assigns them an amount equal to that of all fungi plus all algae. Furthermore, the bacteria are about a tenth of the diameter of algae and fungal. One can show mathematically that for a given mass divided into smaller and smaller spherical packets, the increases in surface area are proportional to the decreases in diameter. Multiplying the factor of two for mass by that of ten for the smaller diameter yields an overall increase of twenty, compared to either algae or fungi. Indeed, using data on the actual carbon density of bacteria, I gauge the global bacterial surface to be about 70 Earth areas.

It could be even higher. Estimates for bacterial populations seem to track upward with each discovery. In the Sargasso Sea, a low-productivity region of the Atlantic Ocean, bacterial counts for the entire column down to 2600 meters imply surface areas for bacterial cells of about 150 square meters per square meter of ocean surface! An upper value for bacterial mass is about five times that used for my first number, producing a planetwide bacterial index of 350 Earth areas. We would expect something impressive from bacteria, and they deliver.

We should briefly note that life has no monopoly on the creation of huge surfaces. Inorganic nature can also make, for example, small spheres. Consider the clouds. Water droplets in these high-floating fogs, summed globally, have surfaces in contact with the air equal to about 100 Earth areas. No wonder sulfates and nitrates are so efficiently scavenged from the air and rained downward.

Another example of big surfaces—this one from the mineral realm—rests beneath our feet. Even the smallest plot of soil harbors a universe of small particles, whose surfaces provide crucial sites for ion storage and release. Here is a shocking calculation: Let's assume that the entire land area is covered with a typical loam of forty percent sand, forty percent silt, and twenty percent clay. Computing over a depth of just ten centimeters (where most of the action of life occurs), and for a moment neglecting the clay, I derive a surface area index equivalent to 2000 Earth areas. Adding the bacteria-sized clay particles boosts this number to 11,000 Earth areas. To calculate a full meter of depth, multiply these numbers by ten.

Despite such splendid displays of surface by inert parts of nature, something special must be attributed to the surfaces of life, something extraordinary. After all, life's surfaces have been honed by evolution because they proved useful for survival. Consider again: green leaves equal to 1 whole Earth area, delicate phytoplankton surfaces of 6 Earth areas, enzyme-secreting fungal hyphae another 6, nearly 40 Earth areas of absorbing root hairs, several hundred Earth areas of furiously active bacterial surfaces. Yes, these numbers pale before the surface area of soil clays. But life specifically evolved these enormous surfaces; they did not have to exist. Life's surfaces are not merely the edges of particles. Life's surfaces boast lipid membranes, primed and powered for the transfer of ions, fluids, gases, or photons. These surfaces facilitate the uptake and regeneration of ten percent of the atmosphere's carbon dioxide each year, the fixation of nitrogen at a rate more than ten times that from lightning, the flow of phosphorus into organisms that is more than twenty times the flux of that element in all the world's rivers. The list could go on. In the surface areas of life, we see the tangible architecture of the global metabolism.

The geometry of living surfaces is thus crucial to gaian dynamics. For example, the extensive living surfaces help resolve a contradiction that you may have noticed buzzing within the logic of this chapter like an annoying mosquito. On the one hand, life has been described as one of four fundamental pools of Gaia and as endowed with rather low levels of some key biochemicals. On the other hand, life lifts its head above the other three and says, "Pay attention to me," because of its role in fashioning the chemical compositions of the others.

Life is alive, of course, whereas the other three primary pools are not. But the surfaces allow us to see something beyond this binary distinction. They suggest an observation about the overall geometry of the relation between life and the other three. Atmosphere, soil, ocean — these are all massive, bulky, and rather coherent. The air coheres as a single volume, mixed by its swirls. Ocean, too, is mostly one volume, mixed to its deepest reaches in a thousand years or so. The soil is not mixed much laterally, but as a blanket between earth and sky, it can house a million nematodes per square meter. It is a substance through which life crawls, grows as roots and underground nets of fungal hyphae, and teems as bacteria in crevices of granules. Within the soil, life is widely distributed and interconnected. In the ocean, too, the food web of phytoplankton, zooplankton, and large life is dispersed throughout the water. Life generally occurs as connected centers of activity within soil and ocean, like the network of spheres and tubes of the Jewish Kabbalah within the uniform background.

Life does not disperse throughout the atmosphere in the way it does in soil or ocean, the flying beings from gnats to bats and the wind-carried spiders and spores notwithstanding. Even so, the unfurling, outward-spreading leaves of plants stirred the Swiss naturalist Charles Bonnet to muse in 1754 that "plants are planted in the air much in the same way as they are planted in the earth." And not only plants. Consider the human lung. As compact animals, we have relatively small outer surfaces. But inside are tucked huge absorbing areas. The exchange surfaces inside lungs, for example, have a total area of more than half a tennis court's worth of membrane for gas exchange—a surface that trades carbon dioxide for oxygen and is thus as active in the air as any tree. Like plants, we are planted in the air. Our exchange surfaces are hidden inside, but they are surfaces in the metabolism of Gaia just the same. Extending this metaphor, one could say that life is planted in soil, air, and ocean.

Because life is literally within the other pools, a distinction may be in order. We need to distinguish life as something special. Let us do so by calling life *life*, while all else becomes *matrix*. This word is related to *mother* and can mean "womb" or, more generally, the "surrounding substance within which something originates, develops, or is contained." That certainly sounds like air, ocean, and soil. These matrixes bathe life, harbor life, surround life, nourish life, and (not least) serve as dump sites for the wastes of life. The three matrixes serve to transfer chemical relations and weave paths of influence among all of life's forms. They surround life because life probes and plants itself into them. Furthermore, because the matrixes themselves are so much products of life, we might embellish the terminology with an adjective. The matrixes might deserve to be called biogenic matrixes. In this we follow the practice of calling trace gases such as dimethyl sulfide "biogenic gases" and rocks such as carbonates "biogenic rocks." But from here on, I will more simply call them the gaian matrixes. The dynamics of Gaia are born of life and the three gaian matrixes.

5
Worldwide Metabolisms

There is no doubt a reclusive aspect to the practice of science: Darwin in retreat at Down House, far from the hubbub of London, patiently studying barnacles; young Newton secluded in the countryside to avoid the plague in Cambridge, contemplating and calculating gravity. Consider as well the anonymous toilers, cloistered for interminable days in laboratories, or (as in my case) in offices hunched over computers. Then, too, there are naturalists, alone in the wild, patiently recording an animal's movement. Every so often, however, the scattered tribe must unite.

For my tribe of earth system scientists, one especially memorable gathering took place in March of 1988 at the edge of the Pacific Ocean, in a sprawling hotel on a tiny island of San Diego Bay. Several hundred of us came together like swarming insects for a great mating dance of ideas. The week-long meeting, sponsored by the American Geophysical Union, has since been recognized as a watershed for Gaia theory.

In addition to the excitement of people and ideas, this conference also afforded me a mini-vision. It occurred just after I had dropped off my bags at the assigned bungalow. Bathed in the California sunshine of early March, I wandered the winding paths of this botanical garden and resort to the edge of a small artificial lake filled with ducks and swans. On the other side of the lovely lake grew a thicket of bushes and trees: palms, eucalyptus, pines, maples, oaks. Within this sumptuous biological playground, anticipating the upcoming meeting, my mind swirled with thoughts and questions about Gaia. What could all these different species mean for the structure of the whole? What role does any individual species play in the planetary realm? Why all this richness?

All of a sudden, I *saw* one answer for the first time. The visual message came from the verdant swath of photosynthesizers on the other side of the lake. It was a message of color: green. The individual trees, shrubs, and grasses blended into a single, sparkling entity. This green beast had contributing parts, to be sure, but these parts seemed more like bricks in a wall than distinct individuals. For the first time, the full impact of a biochemical guild struck me.

Anywhere I go now, the guild might speak to me of its presence. Even here, in the mountains of southwestern New Mexico, although it is winter, many members of the guild have not gone into hiding. Many of those adapted for the inevitable and intense drought of spring and early summer are also adapted for winter's cold: drab camouflage green of juniper in the shade, which turns to a fiery green in the light. Steely green of mountain mahogany, its compact leaves beaming like mirrors in the low sun. Cholla cactus's purple-green stems. Stately long, silvery green needles of ponderosa pine. Scrub oak's green, as if wax were dulled by sandpaper. Cool, white-green of yucca's sword-like leaves. Yellow-green, hair-spike rosettes around the spines of prickly pear: its solar collectors. Gooey strands of algae glowing in a pool, just upstream from emerald patches of puffy watercress thriving in swift icy water. At moments when the universality of the color floods my eyes, a sentence I first heard as I stood before the garden across the lake reverberates in my head: "Gaia cares not that individual species of plants and algae come and go, so long as the presence of green carries forward forever."

Global Green Photon Harvester

Why green? Every schoolchild is taught the lesson behind one of the most sonorous words in science: chlorophyll. One of the major molecules of Gaia, chlorophyll absorbs the red and blue ends of the visible solar spectrum and preferentially reflects some of the middle zone that we call green. Because of the vivid effects that all things that contain it have on light — from two-micron cyanobacteria cells to elephant-ear jungle leaves — chlorophyll is the most in-your-face molecule on this planet. A thick canopy of leaves reflects only a few percent of the intercepted light — mostly green — so the reflected color seems all-important to our eyes. But for the microbe or plant, what is most crucial is that so much of the spectrum is actually absorbed.

Many could recite these accomplishments of chlorophyll. But how many, even within science, would know how to sketch the molecule's shape as something like a squashed tadpole? Or perhaps a flattened bush with a thick taproot? The bush image is a reminder that the distinctive head-and-tail shape of chlorophyll is a matter of function following form: Chlorophyll's head, like that of the bush, receives the energy; its tail, like a root, provides the anchor.

Any molecule is indisputably simpler than any living organism, so perhaps we need a more mechanical analogy for chlorophyll's geometry. Think of a tennis racket. Its flat and relatively round, planar head corresponds to the head of the chlorophyll molecule: Both heads absorb and transfer energy. Unlike the racket's springy mesh, which alters merely the direction of energy (the ball's kinetic energy), chlorophyll's head changes the very form of energy, transforming the photons of light into electron excitation. It accomplishes this with a geometry that is almost crystalline: four pentagonal pyrrole rings contained within a larger porphyrin ring. All told, the four-fold mandala houses roughly thirty atoms each of carbon and hydrogen and three of oxygen, spiced with exactly four of nitrogen (one in each pyrrole ring). A special ingredient sits in the center of the plate: a single atom of magnesium, making its appearance in this book for the first time. This network, primed with loosely bound electrons in layers of rings, can ensnare photons, which it does with varying degrees of efficacy, depending on frequencies.

Chlorophyll's head requires anchoring to properly function as part of the photosynthetic system. This job is performed by the tail. Proportionately more hydrogenated than the head, the tail has roughly forty atoms of hydrogen bristling from a chain of about twenty of carbon (plus two of oxygen). This lipid-like tail sticks fast within the lipid membranes of cells, a case of like attracted to like. Not stiff like the handle of a tennis racket, chlorophyll's tail can bend and twist, rather more like an oily, rolled towel stuck next to similar towels of the membrane's molecules. Finally, to keep chlorophyll from drifting laterally, proteins protruding from the membrane deftly bond to the central magnesium atom of its mandala, ensuring that the photon-absorbing mesh is held at just the right angle.

That's chlorophyll: 120-plus atoms arranged into a binary structure—plate-like head and gooey tail—absorber extraordinaire used by virtually all photosynthesizers. (The exceptions are members of an obscure group of archaea called halobacteria, which use retinal instead of chlorophyll; yes, it is related to the retinal that gives us sight.) Arguably the most important molecule of the biosphere, chlorophyll would be a perfect icon of a science-based, earth-centered religion. In the form of a molecular model, its head and tail might easily replace, for example, the Catholic chalice. A nature priestess could hold the iconic molecule by its tail and lift its illuminated, green head, truly the bringer of light to life, glittering high in the air before a reverent congregation. That might even lure me back to church!

What can we say about chlorophyll's actual distribution in a leaf? Imagine the human population increased by a few ten-millionfold. This multitude, reduced to chlorophyll molecules, would comfortably find niches latched onto internal cellular membranes within a single square centimeter of leaf area. That's the same scale as the density of photons striking the same leaf surface. It enables each chlorophyll molecule to absorb a couple of photons per second from a shower of bright sunlight. Each absorbing plate thus flickers into excited states at about the pace of a heartbeat during brisk aerobic exercise.

The multitudes of absorbers are not randomly scattered, but except in a few cases (for example, in a purple sulfur bacterium), we are not sure exactly how they are arranged. We do know, however, that there are more chlorophyll molecules than there are sites for executing photosynthesis. Chlorophyll molecules are just the first step in a complex process that makes up the integrated photosynthetic system within a cell. Indeed, several hundred chlorophylls typically function together as a multifaceted "antenna" that feeds all its collected energy toward a single site, the reaction center.

Picture a solar power generator in the California desert, with precisely positioned mirrors all around a central tower in which liquid sodium is heated until it glows and develops a thermodynamic power cycle. Each mirror beams its reflected energy directly to the tower. But in the chlorophyll collecting system, each molecule both collects energy on its own and serves as part of an overall bucket brigade, passing along energy that was originally absorbed by others in the antenna. The chlorophyll system can thus be compared to the funnel of a rain gauge. Each unit area on the inner surface of the funnel both collects water directly from the sky and passes along water collected by areas farther away from the central hole. The chlorophyll molecules pass energy from one to the next by excitation transfer, or resonance transfer. One vibrating chlorophyll molecule excites a neighbor and, in so doing, is becalmed. This funneling of activated electron contagion has been likened to energy transfer between adjacent tuning forks.

How the resonance transfer knows that it must move always toward, rather than away from, the reaction center is both fascinating and incompletely resolved. The process is indeed somewhat like water flowing down a funnel. However, the flow is pulled not by gravity but by entropy—a spontaneous flow of energy to lower-energy states, like the conductive flow of heat from a freshly baked loaf of bread. Thus energy from one chlorophyll molecule can jump to a neighbor if the neighbor is at a lower excitation level, which means that its natural pace of vibration is slower. Inequalities in the vibration rates (frequencies) impart the direction for flow. The transfers entail some losses of energy, but the tradeoff provides directionality. In the purple sulfur bacterium, the most closely studied chlorophyll system, chlorophyll molecules are set to three different frequencies, with networks of higher frequencies connected to networks of lower frequencies. Crucial to tuning the frequencies are those adjacent protein molecules that bind to the central magnesium atoms of the chlorophyll molecules and that apparently can control the tension (and hence the frequency) of their sun-catching mandala meshes.

Chlorophyll is helped in its task of collecting and "bucketing" energy from a second class of pigments called carotenoids. The most famous carotenoid, beta carotene, is the dominant pigment in foods such as carrots, pumpkins, and papayas. But beta carotene also occurs in most green leaves, though its presence is masked by chlorophyll. As is apparent by their color, carotenoids absorb light very differently from chlorophyll. Specifically, their blue absorption band is skewed a bit more into the green; without carotenoids, less efficient leaves would sing the blues. Furthermore, carotenoids altogether lack chlorophyll's red absorption band. They thus reflect the "warm" frequencies of light. Depending on the individual types, carotenoids blaze as red, orange, brown, or yellow in autumn foliage, because chlorophyll is the first pigment to fade in a dying leaf. Found in most chlorophyll-containing photosynthesizers, carotenoids add strength to the photosynthetic machinery by collecting photons that chlorophyll would otherwise miss and by participating in the bucket brigades of energy.

Found across the globe in soils, fresh waters, and ocean, cyanobacteria possess chlorophyll but lack carotenoids. They do possess an important third class of pigment molecules. Named after similar molecules found in human bile, the bilins and their associated proteins occur in clusters that fan outward like a bird's tail feathers. The fans are held perpendicular to the photosynthetic membranes, in lines leading toward the reaction centers of photosynthesis, and thus they function to funnel the collected energy, just like the networks of chlorophyll molecules. Curiously, each bilin molecule, like chlorophyll, has four pyrrole rings, but they are positioned in an open chain, as though the closed mandala of chlorophyll had been snipped and stretched linearly. Bilins and chlorophylls, in fact, are close chemical cousins. Their manufacture in the cell follows identical biochemical pathways up to a late point of departure.

The ultimate destinations for the funnels of energy are the reaction centers, tightly integrated units of diverse molecules studded within the membranes internal to the cell. In a reaction center, chlorophyll makes a repeat appearance. Chlorophyll occupies the middle of the reaction center, held by a scaffolding of proteins and touched by a maze of sensitive molecules, each of which can be triggered into accepting and then releasing electrons. This central chlorophyll (which is probably always a uniquely tuned, bonded pair) would be analogous to the hole in the funnel, where the energy is finally whooshed into a different state. When the bucket brigade culminates in resonance transfer to the central chlorophyll, the energy makes an electron leap to a neighboring acceptor molecule. Within the next 200 trillionths of a second, the electron zigs and zags along a maze of acceptors. Operating at speeds like this, the reaction center can easily handle the flow of energy from a collecting array of several hundred chlorophyll molecules. Somewhere in the flurry of activity, as molecules are ratcheted into energized states by trains of leaping electrons, water molecules are cleaved into hydrogen ions, electrons, and oxygen. The end result of the reaction center factory is to shove the negatively charged electrons to one side of the membrane and the positively charged hydrogen ions to the other. (Excess electrons are recycled to chlorophyll to replace those that were lost.) This separation of charge constitutes the official trapping of energy that began as photons and its taming into a form the cell knows well how to manage and use. Remember, the ultimate goal of photosynthesis is to use the energy of light to weld carbon dioxide and the hydrogen from water and then to link these into chains of complex carbohydrates and other molecules. But we are not quite there yet.

Carotenoids also play crucial roles in the water-splitting reaction center. Some help protect the mini-factory against self-inflicted damage. If the pipeline for electrons from the reaction center becomes clogged (which does happen), then the oxygen produced has one of its electrons in a super-excited state. Like a mad dog, it can wreak havoc. Protective carotenoids in the neighborhood absorb the energy from such excited oxygen electrons and harmlessly disperse the energy as heat. (Moral: Don't fool with splitting water without a back-up fire extinguisher.) If all goes well, as it generally must, molecular oxygen flows out of the organism as gas. It may traverse the outermost membrane of a tiny cyanobacterium floating in the ocean's surface, enter the water, and then, via gas exchange, pass into the air. It may exit a cell inside the needle of a pine, enter the air between cells, and then waft out the stomatal pore of the needle and into the atmosphere. The water-splitting reaction center thus links the molecular energetics of chlorophyll to the atmosphere.

In all photosynthesizing organisms that generate oxygen, from microscopic cyanobacteria to towering pines, the electrons from the water that was split travel on to a second and very different type of reaction center. This second assemblage of proteins and special acceptor/donor molecules has its own surrounding array of chlorophyll, which captures more photons and funnels their ensnared energy into a differently tuned chlorophyll pair at the "center of the center." In this second type of reaction center, the energy of the arriving electrons and the additional energy from the funnel are finally shunted into a real, solid chemical compound: nicotinamide adenine dinucleotide phosphate (NADPH). This reaction center might therefore be referred to as the energy-packaging reaction center. In addition, adenosine triphosphate (ATP) is synthesized nearby in another type of molecular factory, which derives power to form the ATP by channeling the hydrogen ions that have traveled from the first, water-splitting reaction center. Together, these two energy-storage molecules (ATP and NADPH) will later be used in the biosynthesis of sugar—crucial for the energy needed for everything else constructed in the photosynthesizing organism.

Now that we have placed chlorophyll in its detailed, functional context, it is worth looking at the global distribution of the molecule. It is not quite as singular as the discussion might have implied. For example, in the needles of a ponderosa pine reside two distinct types of chlorophyll. They are called, conveniently if unpoetically, "a" and "b." It would be difficult to give them more picturesque names, because the minor structural distinction between them is hard to find in a molecular diagram. (See if you can detect the difference in the figure—before reading on.) At one site around the mandala ring, chlorophyll-b sports a side group of CHO, rather than the CH_3 of chlorophyll-a. That's all. These two types of chlorophyll are rather widespread as a coexisting pair. They occur together in all the green plants I live with here in New Mexico and in all the plants across the lake that taught me the lesson of the green biochemical guild. They occur together in all the gymnosperms, all the angiosperms, all ferns, all mosses throughout the world—in other words, in all multicelled plants. They also occur together in some single-celled eukaryotes, such as the so-called "green algae" and the euglenas, and even in a prokaryote called prochloron, which some believe resembles the precursor bacterium that joined the ancestor of all multicelled plants to serve as its chlorophyll-rich chloroplast. All these organisms that contain both chlorophyll-a and chlorophyll-b constitute one tremendous related lobe in the evolution of photosynthesizers.

Many algae, however, do not have chlorophyll-b but instead harbor a different complement to chlorophyll-a. Its name—you may already be one step ahead on this—is chlorophyll-c. Again, comparing molecular diagrams to spot the differences is a challenge in pattern recognition, but only if you see just the heads. Chlorophyll-c has no tail. It is more properly a porphyrin. Bearers of the pair chlorophyll-a and chlorophyll-c include three kinds of single-celled photosynthesizers so important in

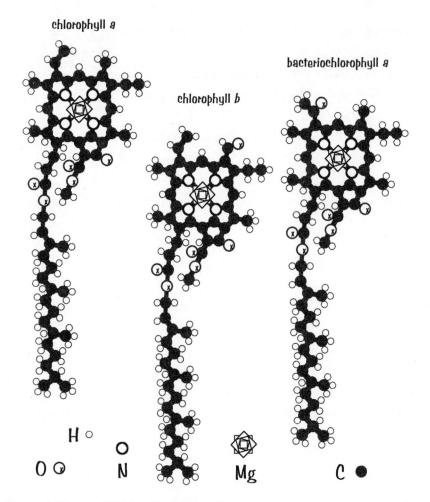

chlorophyll *a*

chlorophyll *b*

bacteriochlorophyll *a*

H ○   O   O ○   N   Mg   C ●

Types of chlorophyll. Molecular diagrams show the stunningly similar detail in three kinds of chlorophyll, with atoms of carbon (C), oxygen (O), hydrogen (H), nitrogen (N), and magnesium (Mg).

the ocean: the coccolithophorids with their white shells of calcium, the golden diatoms with transparent shells of silica, and the dinoflagellates, some of which are responsible for poisonous red tides along coastlines. Sometimes grouped as "brown algae," these possessors of chlorophyll-a and chlorophyll-c constitute the second big lobe in the evolutionary bush of photosynthesizers.

Additional types of chlorophyll are restricted to several groups of photosynthesizing bacteria that do not generate oxygen (the cyanobacteria do generate oxygen, and they possess only chlorophyll-a). In the so-called bacteriochlorophylls, what had been the red absorption band in chlorophyll-a has been shifted to absorb in the far red or infrared. These ultra-red portions of the solar spectrum pass right through the more common chlorophylls of the oxygen generators and are thus available for collection by bacteria that live at the bottom of ponds, or just below the upper surface of mud, or under the surface slime of oxygen-generating microbes in mats. The bacteriochlorophylls come in several types, but again, you'll have to look closely to see the differences between them and the various chlorophylls of the oxygen generators.

Overall, one cannot help but be impressed by the rigid conservatism of the photosynthetic systems. (Think of conservatism as consistency across life's diversity.) Consider the fact that all oxygen-generating photosynthesizers studied so far exhibit the dual dynamo of water-splitting and energy-packaging reaction centers. Consider the near universality of the carotenoid helpers. Consider the ubiquity of chlorophyll, not only in the great diversity of oxygen generators from cyanobacteria to pines, but also in all the anaerobic bacterial photosynthesizers. True, several types of chlorophyll are needed to fill these ranks, but the small differences among them bespeak their close molecular and evolutionary relationships.

Finally, consider a number of points that prove the stunning centrality of the role of chlorophyll-a itself. First, all the oxygen generators possess chlorophyll-a. Second, in the cell's biochemistry, chlorophyll-a is synthesized first, and the other types (even today's bacteriochlorophylls) are derived from it. Chlorophyll-a is therefore probably not just a precursor to the others in the flows of cellular biochemistry, but the precursor in evolution as well. We might now attach a meaning to the "a" — "ancient." Indeed, despite the appeal of the idea that in life's early days, with little gaseous oxygen in the atmosphere, the anaerobic photosynthesizers would have preceded those aerobic ones, many scientists look rather toward the cyanobacteria, with their solo chlorophyll-a, as the evolutionary precursors. Their arrival on the scene may even date back very close to the origin of life itself. Final evidence for the centrality

135

of chlorophyll-a is the fact that it is the terminal goal of the collecting funnel, its pairs enthroned at the center of both types of reaction centers of the oxygen-generating photosynthesizers. In the conservatism of chlorophyll, we therefore see nature following the dictum "Don't mess with a good thing."

The worldwide presence of chlorophyll implies some worldwide genes. It would not be proper to speak of a "green gene," however, because chlorophyll is not a protein. Only proteins, assembled from amino acids, can be considered as directly coded by the genes in DNA. Chlorophyll, rather, is constructed by complex interactions among proteins as enzymes (proteins that snip and clip other molecules, and whose names generally end in –*ase*). For example, one crucial step in the cellular synthesis of chlorophyll is the insertion of a magnesium atom into each mandala ring. This neat trick is executed by the enzyme magnesium chelase (it "chelates" magnesium into some larger molecular structure). One can appropriately speak of a gene for magnesium chelase. Such a gene is therefore worldwide. It varies somewhat across the fields of life, but its universal role in inserting magnesium means that all the varied structures of the enzyme are close cousins. Chlorophyll is constructed by a system of interacting enzymes, and thus behind this green pigment lies not a single gene but a system of global genes. Similarly, we must recognize worldwide families of genes for carotenoids and for some of the other molecules in the reaction centers. The universality of these genes and the enzymes for which they code yields a universal apparatus in Gaia's metabolism, a photon harvester of planetary reach, maintained by the biochemical guild of photosynthesizers.

MADE BY RUBISCO (MASTER MOLECULAR MATCHMAKER)

What about carbon dioxide? The molecule, is, after all, half of the left-hand side of the chemical equation for photosynthesis. Where does it enter the living picture? Thus far we've watched photosynthesis split some water and make some energy-storage molecules during the "photo" part of photosynthesis. Now let's see how it builds the biomass that lifts redwood trees to the sky.

The deep meaning of the "synthesis" part of photosynthesis is the incorporation of carbon from carbon dioxide into organic molecules. Without this crucial first step, the intake of nitrogen, phosphorus, iron, magnesium, or any of the other nutrient elements would be as meaningless as the possession of clothes without a body. Carbon incorporation is accomplished by the "dark reactions," which are so called not because they take place in the dark (although that is possible) but because they are not powered directly by photon energy. Instead, their fuel sources are the energy-storage molecules (NADPH and ATP) produced in the light-driven reaction centers we have already discussed.

First in time and foremost in function in the dark reactions is another universal molecule, the tongue-twisting enzyme ribulose-1,5-bis-phosphate-carboxylase-oxygenase. Most know it more affectionately as Rubisco. It is a giant, a macromolecule, a protein of subunits that are themselves big proteins. It tips the scales at more than 500,000 times the weight of a hydrogen atom. Rubisco is 700 times more massive than chlorophyll. In it, about 50,000 atoms (mostly hydrogen, carbon, oxygen, and nitrogen, the common units of proteins) are bound by means of a holarchy of scales into a big ball, with protruding lobes and folded valleys of shape and charge.

Rubisco is not only big but numerous. It is often hailed as the most abundant individual protein in the biosphere. When you chew a fresh leaf of lettuce, up to half the protein you ingest could be Rubisco.

As its full name indicates, Rubisco is a carboxylase. It therefore promotes carboxylation—the joining of carbon dioxide to something. That something is ribulose-1,5-bisphophate (often known as RuBP). The enzymatic property of Rubisco promotes the alignment of carbon dioxide and RuBP such that their potential mating sites are positioned for rapid and effective union. Like a luxury resort that brings together eligible singles scattered around the world, Rubisco serves as the large context for natural molecular joining. Picture the giant Rubisco embracing both the tiny carbon dioxide molecule and the slightly larger 26-atom RuBP into folds between several of its knobs such that the two smaller molecules snap together. This match wants to proceed. It is a chemical reaction waiting to happen, if only the two reactants can find each other and touch just so. Rubisco crucially serves as matchmaker, catalyzing the introductions and vastly accelerating the rate of coupling.

A divorce follows fast upon the kiss of merger, however. The RuBP-CO_2 system is wobbly and immediately fissions into two molecules of phosphoglycerate (equal dividing of the property). These molecules are stable, so nothing more will happen until yet other enzymes intrude. The phosphoglycerates can then be spun into the so-called Calvin cycle of chemical transformations. Other participants in this cycle include the energy-storage molecules made in the reaction centers of photosynthesis; they are needed to ratchet some of the steps along the way. About midway through the cycle, a spanking new sugar molecule is peeled off. The sugar then proceeds along other cellular pathways that can assemble it into starch and thence into cellulose or a host of other compounds. Some sugars will be burned in respiration to form more energy-storage molecules for the huge variety of synthesis reactions needed within organisms, even for the making of more Rubisco and chlorophyll.

Rubisco is like a Spanish aristocrat with a long string of names, each with a history and meaning. Recall its full name: ribulose-1,5-bisphosphate-carboxylase-oxygenase. The story of its "carboxylase" epithet is the joining of carbon dioxide to RuBP. Analogously, the "oxygenase" moniker must derive from its capacity to catalyze the merging of oxygen and RuBP. This seems to make little sense, because joining with oxygen is usually the doorway to a path of destruction. Catalyzing the oxygenation of RuBP does indeed result in a debit to a photosynthesizer's tally of net carbon gains.

Why would nature invent an enzyme that, like a Hindu deity, first creates and then destroys some of the creation? Perhaps the evolutionary design process that led to a carboxylase gave a geometry that was also, inherently and unavoidably, an oxygenase — a molecular version of no pain, no gain.

Many scientists would likely say unlikely. Rather, the oxygenase activity is a sign of Rubisco's deep evolutionary roots. Photosynthesis probably evolved very early in the history of life, at a time when the atmosphere was much enriched in carbon dioxide and nearly devoid of oxygen. Which of Rubisco's dual enzymatic tugs dominates — carboxylase or oxygenase — is determined by the result of a sort of chemical competition, tugged one way or the other by the concentrations of carbon dioxide and oxygen inside the cell, which are closely pegged to those of the surrounding and interpenetrating atmosphere. Typical estimates of how today's atmosphere affects most plant species are that the oxygenase chews away an astonishing 30 percent of the gains realized from the carboxylase. Experiments with wheat in controlled chambers have established that elevated levels of carbon dioxide promote better growth. That is because the higher concentration of CO_2 boosts Rubisco's carboxylase activity, relative to its oxygenase activity. Interestingly, the same enhanced growth can be had by lowering the oxygen level in the plant growth chambers. This decreases the oxygenase activity relative to that of the carboxylase. Six of one, half a dozen of the other, and the same net result: better wheat growth. If life began in conditions of high carbon dioxide and nearly no free oxygen, then the oxygenase activity of the freshly evolved Rubisco, though potentially present, would have been operationally nil. There would have been no selection pressure to create an enzyme stripped of the oxygenase capability. Once oxygen levels rose, Rubisco was too committed to its role in running the biosphere's intake of carbon to backtrack evolutionarily in fundamental ways.

But there is another theory: Rubisco's oxygenase activity serves photosynthesis somewhat like a coyote's resolve, if caught in a trap, to bite off its foot. If the supply of carbon dioxide fails and the chain of participants in the Calvin cycle is in danger of collapsing, Rubisco's ability to oxygenate RuBP (which leads to both carbon dioxide and phosphoglycerates) can at least temporarily feed an alternative supply of mass into the Calvin cycle factory and keep it spinning. This self-immolation of manufactured molecules in the plant's own respiratory fire seems to help preserve more vital parts of the whole.

Whether an unwelcome leftover from evolution or a crucial safety valve (perhaps it is some of both), the oxygenase activity of Rubisco within today's atmosphere results in a photosynthesis rate that is clearly not maximized. But some forms of life have developed ways to optimize the "atmospheres" within themselves—in other words, the internal environments that surround the operation of their Calvin cycles. For example, many aquatic algae can concentrate dissolved carbon dioxide inside themselves by means of miniature CO_2 pumps in their external membranes. Cyanobacteria, too, generally possess such molecular pumps. The carbon dioxide inside cyanobacteria has been measured at enhancements of a thousandfold. Such intensities nullify the oxygenase activity of Rubisco.

The best-studied and most fascinating method for internally concentrating carbon dioxide exists in at least eighteen different families of land plants: This is the so-called C_4 photosynthetic pathway. When evolutionary convergence of a feature of life is expressed in that many lineages, most if not all of which evolved the feature independently, we can well surmise that an enormous functional value lies at the root of it. Recall from the previous chapter that the BIOME model for research into global change included two types of grasslands. They are distinguished as C_3 and C_4. The "C_3" refers to the regular photosynthetic pathway described above; phosphoglycerate, which carries the carbon dioxide after fission of the unstable $RuBP\text{-}CO_2$ union, has three carbon atoms. In C_4 plants, the carbon dioxide is initially merged as part of a four-carbon molecule. Crops such as maize, sorghum, and sugar cane are C_4 plants.

The genius of the C_4 pathway is that it makes full use of the C_3 Calvin cycle, including Rubisco and RuBP, by means of an embrace that is both physical and biochemical. The C_3 portion is ensconced within "bundle sheath" cells that have been specially enlarged and fortified around the small transport tubules in the depths of the leaf. The multitudes of all other surrounding green cells bind carbon dioxide into a four-carbon compound, which is sent to the bundle sheath cells. There the carbon dioxide is freed by a releasing enzyme and enters the standard C_3 pathway, now confined to the bundle sheath cells. This strategy essentially pumps the concentration of carbon dioxide within the bundle sheath cells to levels up to ten times what they would have been. The chamber experiments with wheat (a C_3 plant) have shown that the oxygenase activity of Rubisco becomes negligible compared to the carboxylase activity when carbon dioxide levels are three to four times those of today's atmosphere. Thus the bundle sheath cells act like miniature chambers within which to run the Calvin cycle at peak efficiency.

James Ehleringer, an ecophysiologist at the University of Utah, has suggested that these enhancements inside the bundle sheath cells approximately record the carbon dioxide level of Earth's past atmosphere around the time of the advent of C_4 photosynthesis. The chamber experiments have shown that the C_4 pathway would have conferred an advantage only at a threshold of atmospheric carbon dioxide concentration a few times higher than today's level. Thus one can search the fossil record for appearance of the C_4 pathway and make a good guess about carbon dioxide levels at that time. But how to search for a *metabolism* in the fossil record?

Because the thick-walled bundle sheath cells show up so prominently under a microscope, fossil leaves can be examined for them. In this case, fortunately, morphology implies metabolism. How speaks the fossil record? C_4 photosynthesis apparently began less than 10 million years ago, rather recently in evolutionary time. Before then, perhaps for several hundred million years, the carbon dioxide level was presumably higher than the threshold for C_4 advantage, and any evolutionary test balloons of the C_4 pathway would have burst.

Why haven't all plants adopted the C_4 strategy? For one thing, the C_4 strategy exacts a metabolic penalty for powering the extra chemistry and transport mechanisms. At very high carbon dioxide levels, C_4 plants would not only be without an advantage, they would be at a disadvantage. Many plant physiologists therefore believe that in a world high in carbon dioxide, C_3 crops such as wheat and rice will do better than the C_4 crops such as maize, sorghum, and sugar cane. Balances would shift in natural plant communities, too. Today, for example, about half the global photosynthesis in natural grasslands comes from C_3 plants and about half from C_4 plants. The C_4 pathway is particularly dominant in ecosystems of hot climates, because the photosynthetic losses in C_3 plants that are attributable to the oxygenase activity of Rubisco rise with temperature.

Let's take a closer look at the eighteen families of flowering plants that have, in at least some of their genera, species that evolved the C_4 function. In addition to the grasses, members of the lily family, the rose family, and the aster family offer examples of this pathway. The C_4 innovation may have evolved independently in each of the eighteen families. Moreover, evolution of C_4 occurred more than once *within* certain of these families—in grasses, for example, probably three times. This is a striking example of parallel evolution, or convergent evolution, whereby genetically distinct taxa have arrived at nearly identical solutions to the same environmental challenges. That wings of bats, birds, and pteranodons all evolved from the forelimbs of tetrapods is an obvious case. More subtle are the similarities between desert plants of America (cacti) and those of Africa (euphorbias). Both exhibit stocky, water-conserving body forms and brandish spines as armor. With the independent evolution of the C_4 pathway, we see an example of what might be called convergent metabolism. It's as though the genomes of many diverse plants that used the C_3 pathway all had the ingredients for some rather complex shifts to the C_4 pathway. All that was needed to induce the change a number of times was a challenging atmosphere.

That the C_4 pathway was an idea virtually poised to be implemented is shown by the fact that even today intermediates exist, lineages somewhere along the evolutionary path to full membership in the C_4 club. Furthermore, there are no fewer than three versions of the pathway that differ in some details. To cite just a single nuance: One version uses malate as the four-carbon transporter of CO_2 to the bundle sheath cells, whereas the other two versions use a different four-carbon molecule, aspartate.

Convergent evolution is the least common way for similarities among taxa to arise. More customary is the principle Darwin made famous: common descent. In the tree metaphor of evolution, that all twigs along a branch of life are mammals — bears, whales, bats, moles — reflects their diverging from the same ancestral branch. The presence of chlorophyll-a in all oxygen-generating photosynthesizers, from cyanobacteria to pines, is almost certainly an example of common descent. So is the respective presence of either chlorophyll-b or chlorophyll-c within the largest two lobes in the evolutionary tree of photosynthesizers.

The general pattern, however, of a second chlorophyll type, whether chlorophyll-b or chlorophyll-c, assisting chlorophyll-a, would be an instance of convergent metabolism. Marshalling a second type of chlorophyll to play a role seems to be an idea whose time came at least twice. Fundamentally different manifestations of the photon-collecting antennae would also exemplify convergent metabolism.

The ubiquity of Rubisco is thought to be the result of common descent. Virtually universal, it toils in all photosynthesizers that generate oxygen and even in some (but not all) bacteria that do not. A complex, gigantic molecule, Rubisco varies in detail from lineage to lineage, unlike the cookie-cutter replication of smaller functional molecules, such as chlorophyll. Mutations in the DNA that codes for Rubisco can slot in a different amino acid here and there. If the mutant's ability to catalyze the joining of carbon dioxide to RuBP is not impaired, then another slight variant of Rubisco may enter the world. If the mutant's catalytic function is improved, then a permanent job for the new variant may be in store. Improving Rubisco offered one direct way of countering the simultaneous rise in oxygen and drop in carbon dioxide that has occurred over Earth's history. As we noted earlier, the oxygenase activity of the enzyme could not simply be "designed away" once Rubisco became ubiquitous and all life depended on it. But small improvements could have been accomplished. When such occurred in distinct lineages of photosynthesizers, the betterment would be an example of convergent metabolism. Descendants of improved forms of Rubisco would then diversify by common descent. Thus the two evolutionary patterns would blend over time. In fact, data from a number of organisms do indicate an improvement in Rubisco's relative attraction for carbon dioxide, compared to that for oxygen, in more recent lineages. The molecular fine-tuning of Rubisco, then, like C_4 photosynthesis, was an idea waiting for the right conditions to happen.

Both common descent and convergent metabolism are ways to spread activities essential to the planetary system. The universal inventory of chlorophyll-a and the presence of C_4 photosynthesis in many plant families are, respectively, exemplars of these two evolutionary patterns. Both lead to worldwide metabolisms. Chlorophyll and Rubisco signal one great biochemical guild, which takes in and transforms carbon dioxide and water, by way of solar power. By their activity in photosynthesis, these worldwide actors bring together not only building-block molecules but also the gaian matrixes of atmosphere, soil, and ocean.

LITTLE ENZYMES THAT RUN THE WORLD

Carbon in carbon dioxide cannot be limited to a one-way transformation, from gaseous into organic form via photosynthesis. That's not the way the world works, not a way it *could* work. Someone or something must recycle the organic carbon back to carbon dioxide. That job is performed primarily by a large guild of respirers.

The global effects of temporary gains in carbon by ecosystems register as a spring and summer photosynthetic drop in atmospheric carbon dioxide concentrations recorded at Mauna Loa. The return flows—via respiration—are just as evident in the annual rise during autumn and winter. The Mauna Loa graph indicates that, at least from the point of view of the atmosphere, respiration is the reverse of photosynthesis. We know this to be true chemically. The reactants and products of photosynthesis are transposed into the products and reactants of respiration. Take the equation for photosynthesis:

carbon dioxide + water + energy → biomass + oxygen.

Reverse the arrow and the result is respiration:

biomass + oxygen → carbon dioxide + water + energy.

Does the reversal extend all the way to the details of molecular machinery? Is respiration perhaps the Calvin cycle run backwards?

Not quite. Nature is not that simple. Let's start with the basics of respiration. Its earliest phase is glycolysis (or the glycolytic pathway). Immediately we have hit upon a universal. Glycolysis occurs in all organisms, regardless of whether they consume oxygen. Glycolysis is the pruning away of some of the chemical energy in sugars. A fermenting yeast, for example, makes alcohol as an end product. The yeast releases the energy stored in sugars that it finds in its environment (say, freshly pressed grape juice) to build energy-storage molecules of its own that it then uses for internal maintenance and to build more yeast biomass. The alcohol is a waste product. Glycolysis does not require oxygen and so is thought to be an ancient biochemical pathway, established in Earth's anaerobic youth. Like Rubisco, this is an instance of common descent, and it is one with even deeper roots—probably from the earliest ancestor of all archaea, bacteria, and eukarya.

Amazingly, some details of glycolysis do indeed look like the Calvin cycle in reverse. For example, one part of the Calvin cycle has the following chemical steps:

phosphoglycerate → bisphosphoglycerate → glyceraldehyde-phosphate.

Now look at some of the early steps of glycolysis:

glyceraldehyde-phosphate → bisphosphoglycerate → phosphoglycerate.

The sequences are mirror images in time. Although some energy-storage molecules must be consumed to drive the steps in the Calvin cycle, the same steps undertaken in reverse during glycolysis *create* those energy-storage molecules. That life can run these and other chemical pathways in reverse is crucial to the dynamics of the grand loops that constitute the biogeochemical cycles of Gaia.

For anaerobic organisms, the chemical endpoint of the glycolytic pathway is the goal. But for organisms such as humans and diatoms and pine trees, the glycolytic pathway is just the beginning. Next comes a cycle. Like the Calvin cycle of photosynthesis, the Krebs cycle is called a cycle because in it, helper molecules are transformed and then ultimately regenerated. For example, because citric acid is utilized as a link and then regenerated, the Krebs cycle is sometimes called the citric acid cycle. Other molecules enter the cycle and then exit when they reach a permanently altered, desired state. The goal of the Krebs cycle, like that of glycolysis, is to derive energy-storage molecules from carbon-containing compounds. In the process, low-energy waste products (carbon dioxide and water) are created and excreted from the cell.

Even though the beginning substances and end products of photosynthesis and aerobic respiration are mirror images, the key inner parts—the Calvin and Krebs cycles—are not. The photosynthetic Calvin cycle is quite unlike the respiratory Krebs cycle (considered in reverse, naturally). One branch of bacteria does, however, run the Krebs cycle in reverse to perform photosynthesis. These are the non–oxygen-generating, green sulfur bacteria. Furthermore, a bacterium called hydrogenbacter, which feeds on hydrogen gas and lives in saline hot springs, can also incorporate carbon dioxide into biomass by a reversed Krebs cycle (in a chemosynthetic, rather than photosynthetic process, because the driving energy is chemical). German biochemist Günter Wächtershäuser has made a strong case that such a reversed Krebs cycle was the primordial means of incorporating the carbon of carbon dioxide into biomass. If this is true, then the machinery for respiration that dominates today's world was inherited from a reversal of this original, synthesizing pathway near the time of life's origins. In any case, the existence of mirror-image pathways alerts us to the importance of rather elaborate biochemical pathways running in both directions within the gaian system.

Universal enzymes make possible the various steps of glycolysis and the Krebs cycle. Universal enzymes make photosynthesis possible. Such species-transcending molecules might be called *the little enzymes that run the world*. In the case of Rubisco and many others, these "little" enzymes are actually giants within the world of lilliputian molecules they catalyze along the pathways. Because the enzymes are invisible to the unaided eye, we most naturally think of them as small. But remember that they can change the air above and the oceans wide in very big ways.

G. Ron Williams, a biochemist and former president of the Canadian Biochemical Society, has probably thought more than anyone else about universal enzymes that run the gaian system. In his book *The Molecular Biology of Gaia*, Williams develops an idea for linking the planetary chemical cycles to the molecular level. According to him, understanding what controls the rates at which molecules are processed by these universal enzymes will be crucial to understanding the grand cycles, because the activities of these enzymes control the fluxes between life and the gaian matrixes. Williams thus points toward a connection between the metabolisms of cells and the metabolism of Gaia.

As an example, Williams has concluded that the nitrogen cycle contains three enzymes of gaian significance. The first, glutamine synthetase, ranks foremost in the pantheon of nitrogen enzymes by virtue of the sheer mass of nitrogen that it channels into the first step in making all the nitrogen-containing compounds essential for life. Specifically, it catalyzes the joining of an ammonium ion (NH_4^+) with the amino acid glutamate, thus making glutamine. The enzyme's name follows from what it does: glutamine synthetase.

Nitrogen exists in a variety of forms: dead organic matter in the soil and ocean, nitrate and ammonium ions in the pore waters of soils and in the ocean, parts of certain minerals, diatomic gas in the atmosphere and ocean, and a bewildering variety of trace gases. Plants require lots of nitrogen, the fourth most massive biological element (after carbon, oxygen, and hydrogen). How is nitrogen incorporated by plants and algae? The universal first step of building nitrogen into the body of an organism requires that nitrogen be present as part of an internal ammonium ion. To attach the ammonium to an organic molecule demands glutamine synthetase. This enzyme is crucial to the biological nitrogen system, somewhat the way Rubisco is crucial to the carbon system, as master matchmaker. From the glutamine formed by the glutamine synthetase reaction, a number of pathways continue on, to create the other amino acids, by transferring the nitrogen-containing amine group from molecule to molecule.

Williams believes the second most important enzyme for the nitrogen needs of life is nitrate reductase. This enzyme is so called because it reduces the positive charge on nitrogen in nitrate (NO_3^-, where N has a charge of $+5$) to a lesser number in nitrite (NO_2^-, where N has a charge of $+3$). Why is this so important? It is the first step in further reduction of nitrogen to the sought-after ammonium ion (NH_4^+, where N has a charge of -3), which can then be catalyzed by glutamine synthetase into glutamine. Because a cell or plant can take in a portion of its nitrogen as ammonium directly, as well as some in the form of nitrate (which must be reduced to ammonium before conversion into glutamine), all nitrogen will ultimately be channeled by glutamine synthetase, but only some fraction of that nitrogen will have been previously channeled by nitrate reductase.

There are reasons why a plant would both prefer ammonium and shun it. Because energy is consumed in reducing nitrate into ammonium, a plant can save resources by taking its nitrogen directly as ammonium. But unless the ammonium is used immediately, it will spontaneously turn into highly toxic ammonia. It is therefore much safer for a plant to take in nitrate, an innocuous and stable dissolved ion. The nitrate can then be transported within the plant's vascular system to stems and leaves, there to be nudged by nitrate reductase on its first step toward becoming an ammonium ion for immediate utilization, right at the site, as needed.

Another issue is the dearth of ammonium in many environments. Ammonium is made by decomposers as they feed on the dead. Then, via other abundant microbes of soil or sea, ammonium can be combined with oxygen to produce energy, leaving nitrate as a waste. Such microbes are so efficient that in many soils and seas, they transform most of the ammonium into nitrate before plants or algae get their chance at it. These activities can be seen in the relative abundances of nitrate and ammonium. For example, the average concentration of nitrate in the ocean is 80 times that of ammonium. In the pore water of typical arable soils, nitrate is 20 to 100 times more abundant. Thus, to the limited extent that plants and algae would "rather" have ammonium, they can't always get what they want. But given nitrate reductase and a bounty of nitrate, they usually get what they need.

The third nitrogen enzyme of global importance is cherished by farmers as well as biochemists. This is nitrogenase. Unlike glutamine synthetase and nitrate reductase, nitrogenase occurs only in prokaryotes, including bacteria and archaea. With it these "fix-it" organisms can take a molecule of atmospheric nitrogen, N_2, break the tough triple bond, and shunt each of the two nitrogen atoms into ammonium ions. Although nitrogenase earns the bronze medal if we judge by the mass of nitrogen that it catalyzes into the cycles of life, its importance cannot be ranked so facilely. Without nitrogenase, denitrifiers would be able to return all nitrogen from fixed forms into the atmosphere as nitrogen gas. They would thus deplete both soil and sea of nitrate, relegating to lightning the sole responsibility for regenerating nitrate. Today, lightning generates only a tenth (or less) as much fixed nitrogen as the nitrogenase-bearing biochemical guild of life.

Free oxygen is anathema to nitrogenase, which breaks down in the presence of even trace amounts of the gas. This is yet another example of an enzyme that evolved before free oxygen became an overwhelming presence in the atmosphere. One of the subunits of nitrogenase is an iron protein with a half-life measured at only 30 to 45 seconds in the presence of oxygen. This presents a serious problem in today's world. Many nitrogen fixers (notably the cyanobacteria) are themselves

149

photosynthesizers and thus produce oxygen. Many others are closely associated with photosynthesizers—for example, the bacteria that inhabit nodules on the roots of legumes, such as soybeans. Photosynthesis in water takes place at the well-lit surface, which is also well oxygenated from intimate contact with the air. Photosynthesis on land usually takes place in well-oxygenated soils. The well-aerated soil that soybean roots require to breathe would spell instant death for the nitrogen fixers.

Thus the nitrogenase-bearing nitrogen fixers must develop ways to ensure oxygen-free local environments in a world flooded with oxygen. Building barriers is one solution. For example, some cyanobacteria form colonial chains of cells, most of which photosynthesize and produce oxygen. But some special cells have been differentiated to exclude chlorophyll and the entire photosynthetic apparatus. Rather, they have thick walls and nitrogenase; isolated from others in the chain, they can safely do the nitrogen fixing. Another strategy employed by some photosynthetic microbes is to fix nitrogen only at night, when they have ceased generating oxygen. Still others, and these are not photosynthesizers, simply rev their respiratory engines to speeds high enough to maintain nearly zero internal oxygen concentrations. Others begin to fix nitrogen only when they experience anaerobic conditions. These different solutions might be considered examples of convergent metabolism: same ends, but varying means. The use to which all these organisms put nitrogenase forces them to create internal anoxic environments in an oxygenated world.

One of the most important solutions occurs in the symbioses of nitrogen-fixing microbes with particular plants. Legumes have been especially well studied. In clusters of bulbous nodules on the legume's roots, grown in response to mutual inducements by the bacteria and the plant, the microbial symbionts respire like mad, which helps maintain low oxygen levels in their bodies. But also crucial for creating local anoxia is a special molecule. Synthesized partly by the plant and partly by the microbes, it is called leghemoglobin, for legume-hemoglobin. Yes, a hemoglobin in a plant, and it is nearly the same as the hemoglobin in our blood! Leghemoglobin soaks up oxygen, which can turn the nodules of legumes pink, and provides a safe harbor for the microbes in the nodules. Like chlorophyll, hemoglobin has a head composed of four pentagonal pyrroles in a larger ring of porphyrin. Whereas chlorophyll has a central magnesium atom, hemoglobin and leghemoglobin have iron. How can microbes, plants, and people all create hemoglobins? It turns out that in the early part of their synthesis, these molecules follow a pathway the same as that for the clearly related chlorophyll and for a wide variety of molecules that assist in pushing electrons to the right places both in photosynthesis and in respiration. In other words, the trunk of the molecular pathway is even more universal than its branches, which lead to a variety of extraordinarily universal molecules of extraordinarily diverse functions.

The world-running enzymes are not always inside organisms. Some are deliberately secreted to catalyze reactions in the near environment. Consider that phosphorus is often in short supply in soils. Most of it is bound into organic materials. Plants need phosphorus in dissolved form, however. When stressed by phosphorus deficiency, many plants secrete the enzyme known as acid phosphatase. Acid phosphatase can attack organic matter and liberate phosphorus into dissolved forms the plant can use. Acid phosphatase is an ectoenzyme; its activity occurs in the soil, outside the plant. It is also an adaptive enzyme, which means the plant controls its release, contingent on the availability of dissolved phosphorus. Similar enzymes are made by many bacteria and fungi, and some of these fungi are attached to roots, again for mutual benefit of symbiont and plant.

Other categories of molecules besides enzymes also can function externally. For example, when legumes are stressed, their roots can release molecules called flavenoids. These molecules diffuse into the surroundings and attract nitrogen-fixing bacteria of the particular genus *Rhizobium*, which live freely in the soil but can also induce and inhabit nodules on roots. In addition to serving as homing calls, the flavenoids throw some genetic switches in the bacteria that prompt the bacteria to secrete substances after they have migrated to the legume and are attached to its roots. These substances alter the plant, and thus begins the cycle of influence whose impact extends to the global atmosphere.

Another class of molecules that are externally released into the soil and there help things run are, like flavenoids, not technically enzymes. Siderophores are released by grasses under conditions of iron deficiency. Iron, like phosphorus, is often problematic to maintain in solution. Siderophores—modified amino acids—exit the cells of roots, diffuse outward, and bind atoms of iron. Being themselves in solution, the siderophores automatically bring the iron into solution as well. Some, with their cargoes of iron, are absorbed back into the root with ground water. These molecules are also adaptive. Their release by the plant declines with an alleviation of stress. As modified amino acids, they possess nitrogen—nitrogen that passed through the sticky cleavage of a glutamine synthetase molecule at some point after entering the plant. Here is one link between the cycles of nitrogen and iron. Such iron-grabbing molecules are secreted by particular bacteria and fungi, which may be a case of ancient common descent or more recent convergence.

Even though all these enzymes and transforming molecules are made and used by organisms, their ubiquity makes it tempting to regard the organism as an unneeded detail. With such a simplification, the cycles of Gaia would be driven not by individual organisms but by biochemical guilds whose members share a commitment to universal enzymes. Ron Williams hypothesizes that learning how cells regulate the activities of these key enzymes will help us discover the rules that build harmony in the global system. We would therefore not need new ecological principles for understanding Gaia; we would need to understand the dynamics of molecular biology (only!) and how to project these dynamics into the global system. In Williams's words,

> It thus appears that the interaction at the biological level of the vast global biogeochemical cycles of carbon and nitrogen may be a reflection of events at the molecular level. One might assert that the regulation of the global biogeochemical cycles is the molecular control on intermediary metabolism writ large. From this perspective,

global metabolism would be seen to be just as much a result of the properties of proteins synthesized under the control of genes as is cellular metabolism. . . .

The global biogeochemical cycles of the nutrient elements do indeed look like the metabolism of a quasi-organism, Gaia. But, perhaps the resemblance arises because these cycles are coordinated and regulated by the molecular mechanisms that underlie the metabolic processes of real organisms. Perhaps we should start to look at these mechanisms with their ecological, even global, significance in mind.

It would be logical to assume that the most crucial molecules of life are those that enable living cells to regulate their internal conditions—those internal molecules that interact with others internally. But the suggestion offered here is that we should not overlook those molecules that catalyze transfers between life and the environment. The three nitrogen enzymes—glutamine synthetase, nitrogen reductase, and nitrogenase—are all key steps on the path of nitrogen flow from the environment into life. So are gatherer molecules released into the soil, such as acid phosphatase and the siderophores. In Rubisco we witness the means by which carbon dioxide is brought under cellular control and guided into organic form.

To the extent that these transfers affect concentrations in the gaian matrixes of atmosphere, soil, and ocean (as they must), the molecular activities influence not only the organisms to which they belong but also other organisms that live within and depend on the matrixes. Such widespread molecules call attention to the biochemical guilds of life: the worldwide metabolisms.

6
Embodied Energy

I n recent years, images of wheat and soybeans growing in giant
chambers on the moon or Mars have claimed a significant share of
my professional attention. I have been part of a futuristic NASA
program that envisions a day when people are permanently extraterrestrial. Like humans everywhere, these explorers and colonists will
need to eat. But the costs of transporting every loaf of bread or sack of
potatoes from the deep gravity well of Earth would be—pardon the
expression—astronomical and thus prohibitive. The astronauts should
be as self-reliant as possible. They should grow their own food.

Human wastes are another concern. These are now simply returned to Earth after each flight of the space shuttle. But in the space-
faring future, the wastes will be recycled as sources of precious elements

that will again become food, and then waste, and then food, around and around in closed loops.

The acronym of this NASA program has always been one of my favorites: CELSS (pronounced "cells"). Originally these letters stood for Closed Ecological Life Support Systems. The program founders soon realized, however, that the material closure referred to in this title could never actually be achieved. On the moon, any doorlock module would leak some gas when opened to the lunar surface to let miners in and out for the day's work. Moreover, perfect closure might not even be the best goal. Mars, for example, has plenty of valuable carbon dioxide in its atmosphere, which could be pumped into the CELSS to grow trees for processing into furniture and plastics, or just for a taste of a green forest from home. Eventually the program managers found a way to keep the felicitous acronym but redo the first word for a more accurate title: Controlled Ecological Life Support Systems.

The acronym CELSS baldly suggests a macroscopic copy of life's own ubiquitous, microscopic control systems. The archetype to emulate in developing this space technology is not the biological cell, however, but the whole Earth. Like the Earth, an extraterrestrial CELSS would have an atmosphere (or several). It would circulate water. And it would have organisms—at the very least, many species of photosynthesizers (the crops), one main species of consumer (the humans), and a bevy of bacterial species (they would be there whether planned or not, so we might as well make them welcome).

My own work has centered on the use of mathematical models to analyze and predict optimal conditions for crop growth and development. How much food can be grown in how constricted a space? What is the most we can hope to achieve in the conversion of energy from photons into the chemical energy of edible plant biomass? To maximize this conversion the growing leaves would be surrounded by an elevated level of carbon dioxide to eliminate the self-consuming, oxygenase activity of Rubisco. The plant roots would be bathed in hydroponic solutions chock full of all essential nutrients in the ideal ratios. Light and temperature, too, would be assiduously controlled and optimized.

With all this pampering, how enticing could the bottom line be, in terms of food production from a given amount of light? We in the program have found it helpful to parcel this analysis into a series of steps, using the conceptual framework of an energy cascade. Think of a stepped cascade of water, which loses some of its potential energy with each step toward the base. In the energy cascade for photosynthesis, a sequence of steps, each with its own pair of potential and actual efficiencies, can be analyzed along the path of transformation from light to biomass: How well (that is, how efficiently) do the crops absorb light? How well do they then transform this absorbed light into simple sugars during photosynthesis? How well do they finally convert the sugars into edible food?

Several chapters ago, we looked in some detail at how the energy input from the sun mixes the atmosphere and oceans. But the flow of solar energy into life itself was given only passing mention. It is time to correct that neglect. The key concept is embodied energy, the portion of solar energy that comes to reside in the bodies of photosynthesizers as *chemical* energy and that is used to fuel the metabolisms of other organisms. The embodiment rate is the amount of energy that actually passes chemically through the biota in its passage from outer space through Earth and back to outer space.

Embodied energy is essential because it gives life the means to power chemical reactions that affect the gaian system. As we will see, the losses of energy between sunlight's entering the atmosphere and the photosynthesis of biomass are enormous. The tool we will use to study these losses—and ask which are inherent and which are due to environmental stresses—is the energy cascade. Because the embodiment rate turns out to be shockingly small, a question arises immediately: How can life exert any significant influence in the Earth system? The answer is that life's embodied energy generates types of mass transformations that either are new or are performed in much smaller quantities in life's absence. An example is the creation of atmospheric oxygen through the burial of only a tiny fraction of organic carbon passed from the engines of photosynthesis.

Moreover, the embodied energy and similar proportions of elements in all living things set up a web of feeding that is fundamental to the chemical cycles of essential elements. These cycles are created not just among organisms themselves but also among organisms and the gaian matrixes. A prominent example is the interlaced pool of decayed organic matter in soil, a store of nutrients that enables the embodiment rate of photosynthesis to be highly amplified. We will look at phosphorus. Furthermore, because life requires elements in certain proportions (always more carbon than phosphorus, for example), and because elements vary in how readily available they are from outside Gaia—from rocks—the cycling ratios of elements also vary. Life's requirements for elements thus affects the magnitudes of fluxes within and among the gaian matrixes and determines which elements will be more and which less under the control of life. Embodied energy underlies our agenda for examining the cycles of all elements as parts of the gaian system.

ᔕ The Energy Cascade from Sun to Plant

Suppose that as you walk down the street, someone slips coins worth a dollar into your pocket. You walk on, and three dimes fall out through a hole in the pocket. There goes thirty cents you never spent and never even really had; it was never part of your financial system. Something quite similar happens to about thirty percent of the solar energy that enters the uppermost, spherical fringe of Earth's atmosphere. Some photons are back-scattered to space by the air itself (the same scattering that bathes the sky in blue). Some photons reach the surface but strike water or snow at such angles that they bounce back upward through the atmosphere and away. The biggest contributor to the total of reflected energy is Earth's reflecting shades—the pearly white clouds so prominent in the satellite images. This reflected energy is like the lost thirty cents: The energy was inside the Earth system in a geometric sense but was never really *inside* the system in an interactive sense.

As we gradually step along the energy cascade to the embodiment rate of photosynthesis—first potential and then actual—the best measure to use is the number of watts per square meter, which expresses a rate of energy flow averaged over a year and over the entire planet. For example, how much sunlight falls on Earth at the top of the atmosphere as primary input? About 340 watts per square meter. This takes into account that half the planet is in darkness at any time and that more energy falls at the tropics than at the high latitudes. The energy flux of almost six 60-watt light bulbs per square meter (bulbs that deliver the full solar spectrum) provides a starting point for what Earth receives from the sun at the top of the atmosphere. Because thirty percent is reflected, 240 watts per square meter is actually absorbed by the various parts of the Earth system.

Not all of these watts are absorbed at ground level. Clouds, for example, are not pure white and thus not perfect mirrors. Their fractal shapes offer numerous nooks and crannies that can trap light and shunt it into heat. Clouds snuff about 15 of the 240 watts. Dust particles aloft in winds, as well as various gases (such as ozone high in the stratosphere and water vapor in the troposphere), consume an even bigger portion — another 55 watts. What is left will all be absorbed by the surface of land and ocean: 170 watts per square meter. Even before it's had a chance to drive photosynthesis, half the incoming solar energy has been lost to reflection and atmospheric absorption.

Let's focus on potential terrestrial photosynthesis, deferring marine until a bit later. In terms of potential, then, we should assume a dense absorbing canopy, such as a thick spruce forest, a tropical rain forest, or even a lush grassland. Although no canopy is perfectly absorbing, experiments with crops such as wheat in the CELSS program's controlled growth chambers have shown that dense canopies of leaves can absorb 98 percent of the types of photons that drive photosynthesis. In any case, for now we will apply no penalty to the energy cascade due to uncovered soil.

The quality of the downpouring light deserves close attention. Not all photons are lively enough to make chlorophyll molecules dance. Sunlight's spectrum of frequencies ranges from those with high energy (high frequencies and short wavelengths) to those with low energy (low frequencies and long wavelengths)—in the language of color, from ultra-violet and blue through green to red and infrared. Only frequencies above a particular minimum (in other words, wavelengths below a certain maximum) can be absorbed by and thus excite the chlorophyll molecules. More detailed discussion of the effects of the spectrum on efficiency will follow. For now, we can consider the crucial frequency to be about 430 trillion cycles per second, which corresponds to a wavelength of 700 nanometers (700 billionths of a meter) and which would appear red to our eyes. Thus we must throw out as worthless all light whose photons are at wavelengths longer than 700 nanometers (basically, those in the infrared). They are too weak to activate the chlorophyll absorbers. What fraction of the solar spectrum must be discarded because of these weakling photons? About 57 percent. The remaining 43 percent constitute the photosynthetically active photons.

A second type of loss attributable to the imperfect quality of photons enters the picture as the absorbed energy of photosynthetically active photons is funneled by the chlorophyll antennae toward reaction centers. Recall that the direction of this funneling is determined by sequentially, in stages as the energy is transferred, constricting the resonant frequencies of the molecules within the funnels. Some energy is unavoidably lost along the way. In the final step, the terminal chlorophyll molecules in the reaction centers are tuned to absorb at about 700 nanometers, or 430 trillion cycles per second, which is where the cutoff point we have cited came from. No matter what the photon's original energy content, here only the portion of that energy that matches the frequency of the central chlorophyll molecules makes the final transfer. What does this mean for efficiency? A precise calculation would require summing over all wavelengths of photosynthetically active photons. But we can make an excellent estimate by simply considering all wavelengths as 550 nanometers, in the middle of the active range from 400 to 700 nanometers. Using this average active photon, we calculate the fraction of energy that shunts into the terminal chlorophyll molecules in the reaction centers; it is 550/700, or 80 percent.

Of the 170 watts per square meter absorbed by an ideal canopy, only 43 percent is photosynthetically active, and 80 percent of that passes into the reaction centers. Multiplying 170 watts per square meter by these percentages in sequence, we find that about 60 watts per square meter enters the reaction centers. With a number of steps to go, one light bulb remains.

The next step in the energy cascade concerns the quantum yield. How many photons (quanta) does it take to emplace a single atom of carbon into a molecule of sugar woven by the loom of the Calvin cycle? Theoretical analyses and detailed experiments have revealed that it takes exactly eight hits of photosynthetically active photons to incorporate one atom of carbon. My colleague in the CELSS project, plant physiologist Bruce Bugbee, believes, however, that the minimum of eight is too few to be taken as the "potentially achievable" minimum for real, whole plants, even those growing in optimized environments. Twelve, he thinks, is more likely to be the minimum. By taking the ratio of the embodied energy of a carbon atom in a photosynthesized sugar molecule with reference to the energy of twelve photons, we can compute the maximum efficiency for quantum yield: about 30 percent.

Before calculating how much energy makes it into the primary sugars, however, we must subtract the loss due to the oxygenase activity of Rubisco—that price paid by all C_3 photosynthesizers in today's suboptimal atmosphere of low carbon dioxide. Allowing for the fact that some regions are dominated by C_4 photosynthesizers, which do not pay this price, and considering the losses estimated from the chamber experiments with C_3 plants, a reasonable global estimate for the inherent loss in today's world from the oxygenase activity of Rubisco is 20 percent. Now let's update the total. Of the 30 percent that "survives" the gauntlet of quantum yield, 80 percent remains after the oxygenase penalty is exacted. The 60 watts in the reaction centers has been reduced to 14 watts per square meter. This is how much, in potential, *could* end up as embodied energy in primary sugars of photosynthesis.

The sugars from photosynthesis enter the networks in leaves and elsewhere both as feedstock for carbon and as fuel. Starches must be made. So must the structural materials of cellulose, hemicellulose, and lignin. So must the thousands of types of proteins. Some sugar is fed into all these assemblies as a source of carbon (and hydrogen). Powering these complex reactions requires fuel. Some sugar is burned with oxygen and respired to drive all the molecular reactions that snip and clip and trim and weld the sum total of the plant biomass. How much of the plant's primary sugars must be oxidized? This varies with the type of plant and its age, as well as with environmental factors such as soil, temperature, and light. But a good, overall, global average is about half. Half of the primary sugars must be burned to fuel the synthesis of all the different molecules and structures that we call a plant. Hence, at this final step in the energy cascade into a fully operational redwood tree or wheat shoot, we must cut those 14 watts per square meter in half, to 7 watts. This is about three percent of the original 240 watts absorbed by the Earth system, or two percent of the 340 watts that fall upon the top of the atmosphere. It is the maximum embodiment rate that can be achieved in today's world with today's plants.

At this point in a similar analysis of the energy cascade in the CELSS program, for crops in growth chambers, the researchers ask themselves, "How are we doing?" Turning the question to Earth, how are plants on real land currently performing, relative to the computed maximum rate by which energy can become embodied, the 7 watts per square meter? The number can be estimated from the energy content of the biomass annually forged by plants. The amount, in terms of carbon, is about 60 billion tons of carbon photosynthesized on land. The total biomass, which includes oxygen, hydrogen, nitrogen, and all the other essential elements, is 2.2 times the amount of carbon, or 132 billions tons per year. For comparison with the maximum potential, this annual biomass must be converted into an energy equivalent: joules per second per square meter (which is watts per square meter). With an energy content of 20 billion joules per ton, an area of 117 million square kilometers (the land area, not counting extreme desert and ice), and the number of seconds in the year, we arrive at the rate of energy flow into the products of photosynthesis: 0.7 watt per square meter. Note that this *actual* amount is only one-tenth of the potentially achievable 7 watts.

To put it another way, the potential flux of energy into photosynthesis is about ten times more than the actual, live-action performance of plants.

Why the difference?

As the saying goes, water the desert and it will grow. Productivities of various ecosystems generally rise as a function of their received rainfall, in a curve of diminishing returns. The wettest ecosystems are about twice as productive as ecosystems that receive the world's average rainfall, about thirty inches per year. Thus watering all the continents to the point of maximizing their potential would lift global productivity by a factor of two. (In some areas productivity would be boosted more; other areas are already saturated.) Water stress thus cuts potential productivity by half.

A second limiting factor is the deep freeze of winter, which affects a portion of the land. Half the land lies within the broad tropical band, where, except in highest mountains, photosynthesis should proceed all year round. What about the other half of the land, which lies above 30° latitude, and occurs mostly in the Northern Hemisphere? Data on the solar energy received over the year in Central Park in New York City show that 60 percent of the year's total energy arrives from May through September. What about farther north? In Fairbanks, Alaska, for example, the growing season might be compressed to three months. A full 48 percent of the year's solar energy arrives from June through August in Fairbanks. Thus, for a quick estimate, allow that half of all land has no seasonal limitation due to temperature and that the other half grows only during 60 percent of the annual light (yes, light, not time). Photosynthesis should thus be active during 80 percent of the total light on a global average.

Some additional loss should probably be included for incomplete absorption. Flying in an airplane on a clear day reveals portions of the country—say Nevada—where the scattered desert plants intercept little of the solar radiation. What fraction of the photosynthetically active photons is actually absorbed by plants at the global scale, from dense forests to scrubland? One difficulty, relative to the value of 100 percent taken earlier as the potential, is that absorption is not independent of water. Neither is it independent of nutrients, a relationship to be discussed next. Scanty rains or sparse nutrients yield low biomasses, and hence a low leaf area index and low absorption. To compute the limitation imposed by real-world absorption, we should figure ideal water and nutrients—and thus a full canopy of trees or cover of grasses. The BIOME model described in Chapter 4 sets the maximum absorption for real vegetation with such high leaf areas at 85 percent.

Finally, nutrients often limit plant production. That soils and the nutrient cycles of ecosystems should ever supply all the essential elements in masses even close to their optimal amounts and proportions seems nothing short of amazing. How can we estimate the limitation imposed by nutrients on a global average basis? One way is to "back in to it."

The limiting factors may be thought of as a series of filters placed in front of a beam of light, where the pure beam represents the global potential for photosynthesis of 7 watts per square meter. The first filter, representing water stress, cuts the light back to 50 percent. The next filter in line, seasonal dormancy, lets through only 80 percent of what it receives. The flux after both of these filters is thus 40 percent of the initial amount, or 2.8 watts. Next comes the filter of 85 percent, for maximum plant absorption. The multiplied total after three filters, then, is 34 percent, or 2.4 watts. Thus the final filter, all nutrient limitations, must let pass through only 30 percent. This is the value that will reduce the cumulated flux, after all four filters, to a final amount of 10 percent, or 0.7 watt per square meter, the number we computed earlier to be the actual embodiment rate of photosynthesis.

The value of 30 percent for nutrient limitation seems reasonable, given the magnitudes of increases achieved with fertilizer on farms. Nutrients are certainly not ideal everywhere. Neither can they be limiting by as much as a factor of ten for a global average, because that would fail to allow for the known effects of the other limiting factors. The calculation here implies that nutrient limitation reduces productivity to about 30 percent of its potential. To put it another way, ideal nutrients would somewhat more than triple productivity. Overall, the four limiting factors, I submit, plausibly explain why actual terrestrial productivity is only about one-tenth of its potential.

Our analysis so far applies to land. But many of the steps identified in the energy cascade also apply to green plankton in the sea. The absorption of light by the atmosphere, the exclusion of the non-photosynthetically active portion from the spectrum, the inherent losses in the transfer of photons to the terminal chlorophyll molecules, the minimum number of photons per embodied carbon atom, the consumption of a fraction of the sugars to synthesize biomass—these are all universal steps for both terrestrial and marine photosynthesis. There would be differences in the actualization of the potential, however. Although absorption can be fairly well quantified for terrestrial vegetation, it is quite difficult to calculate for marine plankton. It would certainly be much lower. Plankton cells can be tumbled up and down in the surface layer. And their densities are often quite low; the ocean is not exactly covered with green pond scum. In some steps of the energy cascade, though, the marine biota fare better. Recall the widespread ability of plankton to concentrate carbon dioxide inside their bodies and thereby overcome the losses to the oxygenase activity of Rubisco. And plankton never lack water. But they often lack nutrients.

To gain a sense of the differences that result between harvesting photons on land versus at sea, let's go straight to the bottom line. How much biomass does the ocean produce? Globally, its output is close to that of land, and I sometimes treat the numbers as equal. But for more precise calculations, most researchers would peg total marine photosynthesis as somewhat smaller. Two-thirds of the terrestrial total is a reasonable number. That would make it 40 billion tons of carbon per year, or 88 billion tons of biomass. It is immediately evident that marine photosynthesis is less efficient than terrestrial: The ocean covers between two and three times the area of land, yet the biomass it produces is smaller.

What happens when land and ocean are combined? What is the global efficiency of the green photon absorber? Over a whole Earth area of 500 million square kilometers, two hundred and twenty billion tons of biomass are produced. (Deserts and the ice of Antarctica and Greenland are now included — it's the fault of the plants that they aren't able to take advantage of those situations.) Following the calculation used earlier for land, we can compute an average wattage entering the embodied energy of photosynthesizers of 0.28 watt per square meter (let's call it 0.3 watt). This is the embodiment rate of energy by global photosynthesis. Compare that to the 240 watts per square meter absorbed by various portions of Earth's atmosphere and surface. The ratio of 0.3 to 240 is just over 0.1 percent. The embodiment rate is 1/23 of the potential of 7 watts per square meter calculated with the energy cascade.

These numbers are far from a rousing cheer for life's big role in the activities of the Earth system. And that's putting it mildly — the case from the viewpoint of energy looks downright pathetic. We could pump up the number by summing the 0.3 watt per square meter of embodied energy over all Earth's area. Globally, that is 150 trillion watts. This sounds impressive, and surely it is. But there is no avoiding the conclusion that the energy of photosynthesis is a trickle to the Niagara Falls of the total absorbed by the Earth system: The value of 240 watts per square meter, summed globally, becomes 120,000 trillion watts continuously entering the planet's various systems.

Where does the 99.9 percent of Gaia's energy dynamics that flows into its nonliving portions go?

The Coupling of Life

> Definition of *edible*: good to eat, and wholesome to digest, as a worm to a toad, a toad to a snake, a snake to a pig, a pig to a man, and a man to a worm."
>
> —Ambrose Bierce

Studies have shown that a quarter of all Earth's absorbed energy drives the water cycle. This amount breaks the sticky bonds between water molecules in liquid and releases the molecules into the air as water vapor. Most of the humidifying vapors arise from the ocean. The soil's surface supplies a third of what lifts from land. The other two-thirds passes skyward through the leaves of plants. Whether it evaporated from sea or soil or transpired from leaves, the water vapor eventually cools and condenses into droplets, releasing its stored, "latent" heat into the air. The most visually dramatic result of this release, on the way toward completion of the water cycle via rain, occurs in the ominous towers of thunderheads, whose volcanic ascent is powered by the molecular explosions along the clouds' columnar cores, from the release of the latent heat of water vapor.

Evaporation and transpiration thus effectively pump energy from the surface into atmospheric heat. Dry convection performs the same trick, those buoyant currents that support eagles and vultures as they soar above cliffs baked by the sun. In addition, because water, soil, and leaves are warmed by solar absorption, winds that pass over them lift their heat into the air as effectively as we cool a cup of tea by blowing across its top. One final method for transferring energy is infrared emission by the generally warmer surface and net absorption by the air. Also recall that some of the solar energy is absorbed directly by the atmosphere.

A small portion of infrared radiation from the surface zips all the way through the atmosphere and away into space. All the otherwise absorbed energy ends up in the atmosphere by one of the possible routes: latent heat, convection, conduction, or radiation. From the atmosphere, the energy (which can never be created or destroyed, only transformed), ultimately flies into space in the only way it can: as infrared emission. Even the embodied energy of photosynthesis eventually escapes. A tree consumed in a forest fire, a grape consumed in our bodies to keep us warm, a leaf consumed by a worm to power movement—all these temporary repositories of life's energy eventually become atmospheric heat and then parts of the spaceward waves of infrared radiation, a flux that maintains the planet's surface at something close to radiation balance. Without the absorption of 120,000 trillion watts of solar energy and its transfer into the thermal heat of air molecules and finally into infrared radiation, the gaian system would be as cold as space. This mind-boggling wattage that is thermally re-radiated to the black void has almost no direct relevance to the embodiment rate by photosynthesis, except, of course, to keep the photosynthesizers warm enough to take in their own, seemingly trivial morsel of energy.

Between its forms as incoming sunlight and outgoing infrared radiation, the energy can reside within and help power the great whirls of atmosphere or ocean. If it were not for this continuous supply of heat, within a week or two the air's gyres would slow to a standstill from internal friction and from the winds grating on the rough and resistant land. Without the atmospheric whirls, the open ocean would become nearly as calm as a windless lake. Tides would still ebb and flow, but even the large-scale thermohaline overturning depends on solar heating, or rather the lack of it at the poles. How much energy must flow into these gyres of gas and liquid to maintain the circulation that unifies the planet and immerses all organisms in a well-stirred bath of air or water? Probably somewhere on the order of 3000 to 4000 trillion watts. This is a couple of dozen times the flow of life's embodied energy, but it is still a small fraction of the total absorbed by the Earth. The circulating whirls are thus an example of how a relatively small amount of energy can nonetheless be vital in operating Gaia.

Let's look at another type of worldwide energy that is even more diminutive than photosynthesis, and one that is not solar-powered. Heat flows from the inner Earth and outward to the surface at about 60 milliwatts per square meter. Globally summed, this energy from Vulcan is 30 trillion watts, only one-fifth of the embodiment rate of photosynthesis. Is this dwarfish amount insignificant to the Earth system? Not unless we can dismiss as insignificant the lifting of mountains and splitting of continents, the venting of gases, the metamorphism of rock, the chemical exchanges at the mid-ocean ridges, and the dynamics of Earth's magnetic field.

These three energy fluxes—global whirls, photosynthesis, and heat-driven tectonics—are all overshadowed by the conversion of sunlight into heat by the Earth system. Yet the whirls and tectonics are both indisputable parts of the system. Life must be, too. But how are we to formulate what the energy of life accomplishes? What is life's genius, if any, in terms of energy conversion? If the swirls and deep heat are any indication, what seems to matter most is not the amount of energy taken in but how it is used.

A clue to life's role in energy conversion can be gleaned by examining one additional type of global flux. This one is much smaller than the 3500 trillion watts of the whirls, smaller than the 150 trillion watts of photosynthesis, even smaller than the 30 trillion watts emanating from Vulcan's deep heat. This flux is the puniest we have considered so far: humanity's seemingly mighty energy engines. Here we add up all the energy flared from the fossil fuels, spun from cascades of water, and winnowed from the nuclei of radioactive atoms. Because most of the total comes from fossil fuels, a quick comparison shows that this total will indeed be less than that achieved by photosynthesis: The carbon waste released as carbon dioxide from all fossil-fuel power conversion is about a tenth of the carbon flux that enters terrestrial photosynthesis. Because carbon transfer is a close proxy for energy conversion, if the terrestrial photosynthetic energy is 90 trillion watts (60 percent of the 150 trillion of global photosynthesis), then humanity would account for 9 trillion watts. Including hydroelectric, nuclear, and the higher energy content (per carbon) of coal explains why the total is somewhat higher, actually 11 trillion watts. This is still only about a third of the flux from Vulcan, which had been the smallest before human technology was weighed in.

What could possibly be worrisome about our release of such a laughably small amount of energy? How could civilization even affect the planet? The answer lies in this well-known maxim: It's not the size but how it's used.

How is it used? To pump carbon dioxide in a one-way trip into the sky. To manufacture chlorofluorocarbons that leak into the stratosphere, where their reactions endanger the integrity of the ozone shield. To pull diatomic nitrogen from the air, break its atomic bonds, and stuff it into ammonium ions for dumping on crops to overcome the nutrient limitations of photosynthesis. To mine ancient phosphorus deposits laid down as sediments a hundred million years ago, again for fertilizer. To plow fields, such that the sediment loads of rivers are now estimated to be twice those that existed prior to agriculture. To blast ore from the ground and smelt it for iron, copper, aluminum. To grind limestone for cement, mix it with gravel, and scrape the sky with concrete towers. To assemble metals into computers and molecules into medicines. To create mountains of waste plastics, substances that didn't exist a hundred years ago. In short, we move matter. And most impressively, we not only move it but perform alchemy upon it. Our genius lies not in the raw amount of energy we use but in how we use it: We transform the chemical states of matter.

Shifting our focus to the natural Earth system, let's apply there the principle that a little energy can mean a lot in the hands of a participant who does something the big players don't. Life, in other words, is like us. Or, rather, we are just life doing what comes naturally to it. Our genius follows the lines of the ancestral genius of life. Talk about the transformation of matter: photosynthesis, glycolysis, respiration, denitrification, sulfate reduction, nitrogen fixation, and the assembly and release of ectoenzymes such as acid phosphates and siderophores.

To be sure, the solar flux without life can blindly drive chemical reactions, particularly in the atmosphere. In the stratosphere, for example, photons can split molecules of water vapor to release hydrogen atoms. The hydrogen can then escape to space, and through a not-so-simple chain of reactions, free oxygen molecules can be generated. The rate of making oxygen in this abiotic way is, however, 700 times slower than the net oxygen flux to the atmosphere that is attributable to the burial of organic carbon made by life.

Buried organic carbon escapes the nearly closed cycle of photosynthesis and respiration and thus does not recombine by respiration with the oxygen that was released during photosynthesis. What fraction of photosynthesized carbon is buried each year? About 1/800 of what is created. Although the burial is mostly marine, we can put this quantity in more familiar terms by considering productivity in a rather healthy—but by no means optimal—region for terrestrial ecosystems. Kenya, Germany, or Paraguay would be enough to supply all the requisite oxygen, were all the carbon produced from any of these regions buried. The resulting oxygen production would sustain Earth's atmospheric pool against the continuous losses during the oxidation of certain minerals, volcanic gases, and ancient carbon in shales. The generative area can be so small because life marshals its embodied energy to drive a very special chemical reaction. The embodiment rate of living energy may be minute on a global basis but its effects shake the planet.

The energy of life is uniquely and effectively applied to drive chemical reactions that would be vanishingly scarce—even nonexistent—without life. Certainly the major transformation is the transfer of carbon from carbon dioxide into carbohydrates and thence into all other types of organic compounds. This act has an important consequence: The embodied energy of photosynthesizers becomes the source of energy for all other organisms. A gazelle on the savanna of Kenya harvests grasses just as those grasses harvested photons. Both gazelle and grass live in fields of energy—the plant in a field of photons, the gazelle in a field of plants. Lions, in turn, live in a field of gazelles, which they stalk as purposively as a gazelle searches for grass, as unremittingly as grass grows toward light. Life is a magnet for other life. And the magnet works dead or alive. Fungi feast in a field of lion dung.

The embodied energy forges a close coupling between forms of life, connects all of life into a single substance, and weaves the intricate patterns of the food web. But there is another reason for the close coupling. The grass to gazelle, the gazelle to lion, the lions' dung to fungi—these are sources not just of energy but also of essential elements, such as carbon, hydrogen, potassium, iron, and other building blocks of bodies. The fungus can absorb minerals from the soil as well, and sometimes a lion may swallow some dirt in search of salt, but the bulk of what is ingested by the non-photosynthesizers comes straight from other life. Furthermore, in all these food sources, the elements are correctly proportioned for the next in line.

Compare the composition of a grape leaf to that of a human. The top four elements in both, in order of descending mass, are carbon, oxygen, hydrogen, and nitrogen. True, there are differences between leaf and person. (The percentages of the top four, for example, differ because the quantities of proteins and fats in humans boost the fractions of carbon and nitrogen. Potassium claims fifth place in the leaf, but

Leaf and human. The relative areas of the circles show the abundances of elements in both plants and humans. From left to right, the elements are as follows (the numbers in parentheses give the percentage of the dry biomass that each element represents, first for the leaf, then for the human): oxygen (45.2, 15.2); hydrogen (6.0, 9.0); nitrogen (1.6, 7.7); sulfur (0.1, 0.4); phosphorus (0.2, 2.5); potassium (1.0, 0.5); calcium (0.5, 4.3); magnesium (0.2, 0.1); iron (0.01, 0.01); carbon (45.2, 60.3). Several micronutrients are not shown.

calcium fills that rank in the human because of its presence in bone.) Yet behold the overall, striking similarities. The micronutrient iron, for example, is about 0.01 percent in both.

Now consider that food provides energy for metabolism and movement as well as materials for body building and repair. The proportions of elements required by a person for these combined tasks shift toward the "plant end" of the spectrum. The human diet is usually higher in carbohydrates, and proportionally lower in fat and protein, than the human body. But delving into such detail is beside the point. As different as human and leaf seem to be in color, form, texture, and habits, they are remarkably similar chemically. A plant on a plate offers you something quite close to what you need: mostly carbon, for example, and a tiny bit of copper. You obtain, fortunately, more nitrogen than phosphorus, more phosphorus than iron. Imagine the complications if our bodies were half magnesium, rather than half carbon, and were trying to grow or just renew themselves by eating plants that are less than one percent magnesium. We would starve digesting all that mass to filter out what we required. Instead, because of the excellent match, we need think very little about what we eat. We can just shovel it in and know that our bodies will do pretty well. It is worth repeating: The proportions of the elements in a human body—and in all animals, fungi, and bacteria—are remarkably close to those within a forest. The match is not by chance. A common ancestor and the selective forces of evolution are behind the chemical unity of life.

But why these particular proportions of elements? Carbon is there primarily because it provides the molecular frameworks for other atoms, mainly hydrogen and oxygen. (Hydrogen atoms are the most abundant in life, but hydrogen is only third in mass.) Nitrogen is needed for all amino acids of proteins, and sulfur for two of the twenty-odd amino acids. Phosphorus is vital in many molecules that store and release energy and in some carbohydrates active in cell metabolism. Single atoms of magnesium, as we saw, occupy the center of the mandalas of the chlorophyll molecules. The chlorophyll molecule reflects life itself fairly well: a lot of carbon and hydrogen with a pinch of special ingredients.

These proportions have consequences for global physiology: The fluxes in and out of life are closely tied to the proportions needed internally for the compositions of organisms. This is most obvious when one organism feeds on another. It is also true for those inevitable times in the global biogeochemical cycles when the elements leave life and spread out into the gaian matrixes of soil, air, and ocean, and for those times when the reverse occurs, and these elements are pulled from the matrixes, primarily by plants and green plankton. For example, because the concentration of magnesium in a plant is so much less than that of carbon, the flux of magnesium ions from the soil into plant roots is much less than the net flux of carbon from carbon dioxide into the leaves during photosynthesis. The physiology of organisms affects geophysiology by determining the ratios of fluxes of various elements between life and environment. Thus, globally, carbon's flux between the gaian matrixes and life is greater than nitrogen's, and nitrogen's flux is greater than that of phosphorus, which in turn is greater than iron's.

We must be careful not to equate rank in terms of mass flux with importance. A collapse in the supply of magnesium would bring photosynthesis to a halt just as surely as depletion of carbon dioxide from the air. That is why they are *all* called essential elements. Furthermore, what matters is not the size but how it's used. Small fluxes can have big effects, as we now see from the trace amounts of chlorofluorocarbons we have unleashed upon the ozone layer. The impact of a flux depends on the form of the matter in the flux and on what it chemically or energetically reacts with in the environment. Nonetheless, the flows will be coupled to each other on the global scale in proportions that roughly reflect their abundances within life.

As an example, carbon moves into and out of photosynthesizers in fluxes of about 100 billion tons per year. Nitrogen's analogous fluxes are less, about 8 billion tons per year; this reflects the way organisms are built. Life has—and thus needs—less nitrogen than carbon. (Differences between marine and terrestrial fractions are discussed in the next chapter.) Nitrogen fixation by the special microbes totes up about 150 million tons per year, a figure inherently lower than the total nitrogen fluxes because this process brings only new, fixed nitrogen to the terrestrial or marine systems. Note the order: The total carbon flux is more than the total nitrogen flux, which is more than the nitrogen fixation flux.

Ron Williams, as we saw in the Chapter 5, has proposed that a key understanding of Gaia will come from understanding the controls on the fluxes of elements to and from life. A clue to those controls may be found in the fact that life is organized such that the coupling of fluxes entering and leaving corresponds to the ratios, within life, of elements needed for growth and replacement of parts. For example, the adaptive ectoenzyme acid phosphatase, which can release phosphorus from organic molecules in the soil, is not made by roots when phosphorus is abundant. Why waste energy and materials releasing more phosphorus than can be used? In a nutshell, the little enzymes that run the world create flows in proportions that are required to make more little enzymes.

This matching of effort with need requires a cross-talk of feedback and close coupling between the composition of life and its fluxes of matter in and out. What it takes to run life implies certain proportions of element fluxes. And those proportions of element fluxes build the molecules that actually cause the fluxes. Neat. The logic is cyclical, which is appropriate because life itself is above all cyclical. I suspect that herein lies a far-reaching pointer for thinking about the physiology of Earth.

The Imprint of Life

The experimental crops grown in the CELSS program are pampered in every way. We give them cloudless "skies," perfect temperatures, nutrients galore, and a drought-free, pest-free, carefree life. We root them not in soil but in a shallow, nutrient-laden bath called a hydroponic solution. The plants respond to the gentle, nonabrasive fluid by putting forth a modest bunch of roots, perhaps one-fifth of the root mass spread by a crop in the field.

Central to keeping the hydroponic roots happy and thus small is an ideal mix of all dozen-plus essential nutrients. The most abundant in the hydroponic solution is nitrogen, usually as a mix of nitrate and ammonium ions. The least abundant is inevitably molybdenum, a catalytic cornerstone in nitrate reductase, the second most active enzyme in the nitrogen cycle. A good hydroponic solution contains nitrogen and molybdenum in a ratio of about a million to one, which is near their ratio in the growing biomass.

Maple trees in a Vermont forest and grasses on the Kenyan savanna have no plant physiologists daily monitoring their nutrients and ready to supply refills as necessary. Plants coping with the vagaries of the real world must rely on a natural balance of ions in the waters of their soil's pore spaces. The balance is often not perfect. As we noted earlier, stresses from inadequate nutrients currently restrict terrestrial photosynthesis—perhaps by an average factor of three. And yet forests often do grow tall and grasses thick. Though not coddled by plant physiologists, plants in nature do have a matrix of other life forms, the soil, and even themselves to depend on.

When plants die or shed their parts, the falling material carries earthward the essential elements in their living proportions. Perhaps ninety percent of all plant biomass meets this fate (note that roots are already in the soil when a plant dies). The nutrients in the other ten percent, that fraction of annual photosynthesis eaten by insects and other animals, eventually also return to earth as animal wastes and corpses. All this detritus harbors some embodied energy, for even animal wastes are far from fully oxidized. The energy, along with a congenial ratio of elements, makes the detrital amalgam a tempting home for nematodes, fungi, and bacteria—and thus a catalog of strange soil organisms. As these denizens of the underworld extract materials for their tiny bodies, and these organisms in turn are consumed either alive or dead by still others, the material passes from one little mouth (or permeable membrane) to another. The chemically stored energy, however, inevitably degrades, as the materials move closer and closer to disintegrating into isolated ions in the soil. As the embodied energy of the

Life from life. This bird's nest fungus, growing from a fallen chunk of wood, takes the wood's elements and those from dead bacteria into its own body and thereby illustrates the close coupling of all life. The nutrient-gathering hyphal threads of the fungus are dispersed within the wood. We see here only the mushrooms. The "lids" are off on several, exposing the spore cases within.

biomass moves and undergoes transformation along the strands of the soil's food web, the form of the matter becomes more "to the liking" of a plant. Because plants harvest solar photons for energy, they do not require energy in the matter they take in. In fact, the nutrient matter for plants must generally be in a low-energy condition — in other words, as dissolved, inorganic ions — to be available at all.

How effective is the closure of the soil nutrient cycles? Like a NASA-built CELSS, the cycles appear to be controlled to achieve a great deal of closure. Yet some nutrients always leak away. Dissolved in water, they may percolate down beyond the reach of roots, enter ground water, and end up in streams merging into rivers and whooshed to the ocean. Though such leaks inevitably occur for all elements, it is not at all obvious which of the essential elements suffer the greatest leaks. For any essential element, what is the ratio of the flow within its ecological cycle to the flow of its leak? In Chapter 2, this ratio was dubbed the cycling ratio. Knowing how many times, on average, life turns an atom of an element around within its own system of control before that atom is lost from the system may reveal something about the imprint of life on the very cycles that life participates in.

As we prepare to discuss the soil's nutrient cycles in more detail, a milkshake is in order. Suppose you are waiting in line at Dairy Queen. Having nothing better to do, you watch the server set a small container under the vanilla machine and turn it on. She also sets a large container under the chocolate machine and flicks it on, but then she is distracted by an emergency at the french-fry maker and forgets the milkshakes. This enables you to watch both containers overflow. In the few seconds before the server returns and shrieks, you have noted a few general principles. First, both shakes are in steady state, a condition in which the inflow and outflow are equal. In fact, by directing the overflow into a channel and then — no, not into your mouth — into a graduated cylinder borrowed from a chemistry lab, you could accurately infer the inflow by simply measuring the outflow. Furthermore, the server just happened to toggle the controls in such a way that a visibly larger stream transited through the smaller, vanilla container. Thus you note that once the system is in steady state, the container's size does not matter: the smaller container in this case supports the larger outflow.

With a conceptual straw filled with the overflowing milkshakes, we are ready to imbibe the question of how the soil cycle works for different elements. The first element to be examined is phosphorus. Phosphorus is often in short supply, which is why it is included in many fertilizers and is commonly cited as a crucial control on productivity on land and at sea. Its lack of a gaseous phase makes its cycle relatively simple.

As a component of soil's dark, humic organic matter, phosphorus totals globally about 10 billion metric tons. That is impressive: it is four to five times the amount in the world's vegetation. And yet this figure pales before the amount of mineral phosphorus bound in soil's particles:

about 100 billion tons in the topmost meter. To put this number in context, consider that compared to the other elements in soil minerals, phosphorus is rather scarce, constituting only about one-tenth of 1 percent (less in sedimentary rock, more in igneous rock). A dozen mineral elements are more abundant than phosphorus, including other biological essentials such as calcium, magnesium, potassium, and iron. Nonetheless, this piddling amount of mineral phosphorus is ten times more than the phosphorus in the soil's organic matter.

What about the fluxes into and out of these two soil phosphorus pools? Recall the lesson of the overflowing milkshake. Just because the mineral pool is larger does not mean it is overflowing (er, being depleted) and being resupplied at a greater rate than the organic pool. In fact, we know that the detritus from life provides a torrent of organic phosphorus to the soil: an annual flux of about 230 million tons, enough to refill the organic pool in about forty years. That this organic pool remains about the same size means it is nearly in steady state, so its exiting flux is also 230 million tons a year.

What are the fluxes entering and leaving the soil's mineral pool? The source of net phosphorus is the nether bedrock of Vulcan. We have no direct way to measure the physical weathering of rock. It must be computed indirectly, via the principle of the overflowing milkshake.

The exiting flux from the mineral pool, unfortunately, is also not directly measurable. But what leaves the soil's network of pools is precisely funneled into the world's rivers, and researchers have been able to measure that. Rivers contain atoms of phosphorus packaged in three different ways: as dissolved ions, organic particles, and mineral particles. Perfect for plant growth, the dissolved ions count as a definite loss of nutrients from the system—about a million tons of phosphorus per year going, going, gone into the oceans. Organic particles, the second kind of package, consist of microscopic bits of phosphorus-containing organic matter attached to clay particles carried by flowing water. The pre-agricultural estimate for these, prior to the increase from human-caused erosion, is 4 million tons of phosphorus per year. Finally, phosphorus bonded with calcium, iron, or other inorganic elements in mineral particles streams toward the sea in an annual amount of 6 million tons.

These three distinct phases of phosphorus in rivers—dissolved ions, organic particles, and mineral particles—represent a loss from the soil system that can ultimately be replenished only by a flux from the bedrock of Vulcan. Invoking our overflowing milkshake, we can sum the three fluxes to obtain the flux from Vulcan: 11 million tons annually. To keep the picture simple, but still accurate for our purposes, let's assume that these 11 million tons go first into the soil's mineral pool.

Out of the 11 million tons that enter the soil's mineral pool, 6 million tons leave as the mineral particle flux in rivers. The other 5 million tons must go somewhere (again, the overflowing milkshake). They go into one additional pool, whose global phosphorus content I would not presume to tally. It is small and often deficient in phosphorus. It is the pool of dissolved phosphorus ions in the soil water itself. The 230 million tons per year that plants must suck from this pool (to produce the detritus flux to the soil) clearly cannot be refilled by the 5 million tons coming from soil minerals. The difference must be made up by the soil's organic pool, by organic molecules digested and excreted, primarily by microbes, into the soil water as dissolved ions.

I have simplified what is a much more complex and imperfectly understood network of internal pools and fluxes of phosphorus within the soil. (Such simplified, homogeneous pools are extraordinarily useful as analytical tools.) For instance, phosphorus ions from the soil water attach to clay minerals, becoming part of the mineral pool, and then detach again with varying degrees of efficiency. But our simple picture deals with global totals of net fluxes and is based on measurable quantities such as the fluxes to and from photosynthesis and that of rivers. Thus in the diagram, all the pools (think milkshakes) can be balanced by using these "measureables" to derive other fluxes within the system. This balancing enables us, finally, to assign a number to the efficiency of phosphorus cycling between plants and soil.

The riverine losses that most directly affect the annual requirement of photosynthesis for 230 million tons of phosphorus are the 4 million

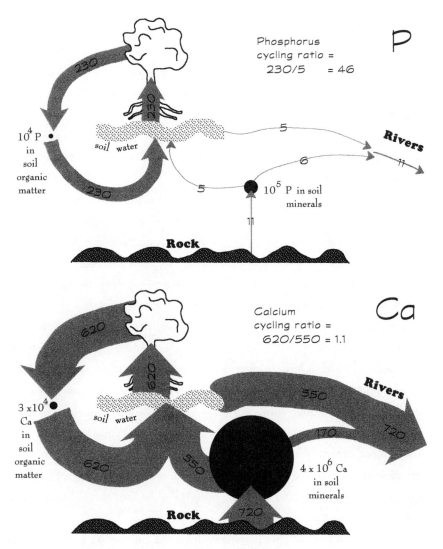

Terrestrial cycling ratios. Phosphorus, with a high cycling ratio, and calcium, with a low cycling ratio, exemplify extremes among the elements essential for life. The widths of the arrows show the fluxes (in millions of tons per year), and the areas of the circles show pool sizes (in millions of tons). Note how large is the cycle of phosphorus from vegetation counterclockwise and back to vegetation, compared to the phosphorus fluxes entering from soil minerals and exiting by rivers. In contrast, all the fluxes for calcium are about the same size. The availability of phosphorus often limits plant growth, because its supply from rock is so low relative to the needs of plants.

tons in organic particles and the million tons in ions from the organic soil pool and the soil water pool, respectively. These account for a total loss of 5 million tons. Note that we are not counting the 6 million tons that enter rivers as particles from the soil mineral pool, because this form of phosphorus is neither available to plants nor a product of them. These particles have never entered the cycle of life. They skirt the pulsing activity of the biological district like trucks looping around the freeway bypass of a city.

Now let's compute the cycling ratio. For every 230 atoms of phosphorus spinning around in the nutrient cycle between plants and soil, an average of 5 will exit, stage river. The cycling ratio is thus 230/5, or 46. A single atom of phosphorus will spin around between plants and the soil's organic matter an average of 46 times before being lost, via rivers, as either an organic particle or a dissolved ion.

Another way to look at this cycling ratio is to consider the net input to the nutrient cycle from the soil's minerals, proximately, and from bedrock, ultimately. Were the bodies of plants not transferred into the soil's organic matrix and then recycled into themselves, they would have to rely for photosynthesis on the 5 million tons from the soil minerals alone. Photosynthesis would be cut by a factor of nearly 50 (OK, 46). Instead, through a fortuitous pair of life's virtues—its embodied energy and its congenial proportions of elements—the system of biology and soil matrix amplifies the availability of phosphorus, and thus photosynthesis, by a factor of 46.

The cycling ratio thus provides the same numerical answer to either of two questions: How many times is an element recycled in and out of photosynthetic life before it is leaked from the cycling system? And how much photosynthetic amplification does the cycling system provide to the flux of an element supplied to it? For phosphorus in the terrestrial cycle, the answer is 46.

What about calcium? Calcium is found throughout living cells as a regulatory ion in control systems; plays another vital role as a structural element in plant cell walls, as a bridge between certain molecules called pectins; and also has a structural role in animal bodies. The flow of calcium into terrestrial photosynthesis is somewhat larger than that of phosphorus—about 620 million tons of calcium per year. Calcium is extraordinarily abundant in soil minerals (40 times more so than phosphorus); thus there is about 4000 billion tons of it. This abundance is reflected in its river flux of 720 million tons per year, which is somewhat more than 40 times the river flux of phosphorus but not out of the ballpark. The river flux of calcium is distributed between particulate and dissolved forms in a ratio entirely different from that of the relatively insoluble phosphorus. About 550 million tons of calcium is in the form of dissolved ions and 170 million tons in mineral particulates. Again, I have estimated these as values prior to agriculture, logging, and industry.

We now have the information for calcium that we need to apply the reasoning used earlier to derive the relevant fluxes for phosphorus, one by one, by sequentially invoking the overflowing milkshake principle. The calcium diagram shows the detailed results. The bottom line comes from the following key numbers: The organic cycle among plants, detritus, and soil water flows at a rate of 620 million tons of calcium per year. The river loss of potentially available calcium ions from the soil water is 550 million tons per year. The ratio between photosynthetic need and this loss is 620/550, so the cycling ratio is about 1.1. Were photosynthesis limited to a single pass of this dissolved flux that comes from soil minerals, enters soil water, and leaves by rivers, life could incorporate 550 million tons per year, or about ninety percent of what it currently uses. The biological amplifier in the case of calcium is thus only ten percent.

The cycling ratio for phosphorus is 46. For calcium it's barely above 1. That's a radical difference in how these elements behave in the terrestrial gaian system. Life only slightly enhances its productivity by recycling calcium to supplement that which enters the soil water from the dissolution of soil minerals alone. On the other hand, life greatly amplifies the availability of phosphorus. The difference comes from the combination of physical and biological factors. These factors are unique to each element, and the diagrams and numbers show the results. But what does it all mean?

It means that if you're a farmer, you are more likely to be thinking about putting phosphorus on your fields than calcium. Phosphorus, being more dependent on a nearly closed, carefully controlled cycle between life and the soil matrix, is more likely to become deficient because it lacks the available back-up system of abundant supply that calcium has courtesy of Vulcan. It turns out that potassium also has a fairly high cycling ratio. Not surprisingly, potassium (symbol: K) is often an ingredient in fertilizer, along with phosphorus. A third element that is most universal of all in fertilizer is nitrogen, in the standard N–P–K triad. The cycling ratio of nitrogen, if calculated the same way as for the phosphorus and calcium diagrams, is also rather high. But nitrogen's fluxes to and from the atmosphere in nitrogen fixation and denitrification complicate the picture. Nonetheless, we see in phosphorus, potassium, and nitrogen a concordance between microcosm and macrocosm. Down on the farm, the most deficient element limits overall production and thus must be supplied first, a principle known as Liebig's Law of the Minimum. In the terrestrial cycling ratios, we see this law writ large—magnified to the global scale.

This is not to imply that calcium can always be ignored. A shot of calcium can help control acidity, for example. But all in all, the most globally deficient, sensitive, and susceptible elements are those with the highest cycling ratios on the large scale, on the scale of the major gaian subsystem between life and soil. After all, why is it that life, by way of evolution, bothered to bring about such high values in the first place? Answer: Vulcan was a stingy supplier, so life simply had to recycle. The local needs of farms are perfectly reflected in the global numbers, based on the relative needs of terrestrial photosynthesizers for essential elements and the fluxes of these elements in rivers, which correspond to their availability from the rocks of Vulcan.

Other than grounding the concerns of farmers in the deeper level of causation arising from the relationship between Gaia and Vulcan, what, if anything, do the different cycling ratios mean? At least three things: Each element has its own story in geophysiology; the elements with the largest cycling ratios are the most crucial to understand; and the elements with the largest cycling ratios provide an indication of how well life is doing relative to its achievable potential.

First, all the essential elements possess unique behaviors that must be understood to paint a realistic portrait of Gaia as a system. Each element has its own individual proportions within life, within soil organic matter, within soil minerals, and within rivers. Some elements, furthermore, enter and leave as gaseous forms, thus adding new types of fluxes to their diagrams. Somehow they all are consolidated into an interlocking network, traveling sometimes together, sometimes separately.

Next, those who would undertake a study of geophysiology should look first at those elements with the highest cycling ratios. Deficits of these critical elements are most likely to jeopardize an ecosystem, because Vulcan is not forthcoming with replacements. Life can safely squander elements that are easily available for free from rocks. But the elements with the higher cycling ratios are those to which evolution would have devoted the most attention — the cutting edge where new discoveries in metabolism could create major advantages and boost the overall productivity of life.

Finally, how to estimate the power of life? Because improvements in sharing an element with a high cycling ratio within the system of life drive the system toward greater productivity, we would use that element for understanding the relationship between the actual and the possible. As described earlier, nutrient limitations probably reduce terrestrial photosynthesis to about thirty percent of what it would be with ideal nutrients. Thus from a nutrient perspective, life is about one-third of the way toward achieving the possible, and phosphorus, with the highest terrestrial cycling ratio, is the greatest limiting factor.

But there is another way to look at phosphorus in evaluating the achievement of life on Earth. If life is currently amplified by a factor of roughly fifty, and if life could only be boosted an additional factor of three by even more effective recycling before running into the other limitations of water, temperature, and light absorption, then it would appear that life is much more than thirty percent of the way toward its potential, measured not linearly but as multiples of a cycling ratio.

Suppose we start with life having to rely solely on the annual flux from Vulcan, an irreducible baseline with no recycling—a cycling ratio of 1. Now consider that each step on the path toward more and more recycling accomplishes some given reduction in the fraction that leaks from the detritus. For example, step number one might cut back the leak until it was only one-third of the detritus flux and return two-thirds toward the plants. Productivity would rise until the original amount of leak was restored, because in the steady state, it must be equal to the flux from Vulcan. This would result in a productivity three times the initial rate, so the cycling ratio would have grown from 1 to 3. Say step number two constricts the leaking fraction of detritus again to a third; productivity thus rises to nine times the original rate to restore again the steady-state condition with respect to Vulcan. The cycling ratio is now 9. In step number three, the leak rate from detritus is further reduced to a third, and by the same logic, the cycling ratio rises to 27. For the sake of easy argument, consider this figure close to the value of 46 for the cycling ratio of phosphorus. Now, assuming that nutrients could triple the productivity at most once more, the system has only one more step to go—that is, to cut the leak to a third again and thereby triple the cycling ratio. Here we stop because water, temperature, and light absorption now become the barriers to higher productivity. This analysis suggests that the nutrient-cycling capabilities of life have already accomplished three out of four possible steps. Thus life has achieved not 30 but 75 percent of the potential. Bravo life!

How does life enhance its ability to recycle? We really cannot speak of life as a whole evolving. Rather, we must focus on the evolution of organisms. Soil organisms, which perform the recycling, do not recycle for the benefit of plants. But as they consume detritus for their own needs, they inadvertently recycle by keeping in circulation material with a congenial proportion of elements. Their stripping of the embodied energy from the detritus heads the detritus toward exactly the state the plants need it to be in to spin another round of nutrient control. The cycling ratio of the whole changes as life forms evolve to fill the ecological niches available upon the various types of detritus in different degrees of degradation. By existing in dense variety, living things in the detritus-based food web together help close the loop, enabling all to thrive better in their controlled ecological life support system.

7
The Music
of This Sphere

I t has snowed for the past two days. Flurries drifted down from
white skies, endlessly. Finally today the blue broke through. The
sun heated the rocks that poked up through the snow like little
furnaces, melting the white away in rings, exposing soil to absorb still
more sun. The tiniest arroyos have begun to run with riffling, gurgling
rivulets.

Refreshing river experiences! A rich way to recall my journey
along the river of life is by the rivers I have known: the Niagara, where
I grew up, virtually a moving lake; then the mystical Willowemoc in the
Catskills of upstate New York; later the majestically industrial Hudson
of New York City. Punctuating the trip have been Flat Creek in the
misty mountains of North Carolina, where I could walk a mile along

the middle of an ankle-deep stream until the roar just around a bend signaled a hundred-foot plunge; Siberia's Yenisey, whose tumultuous, Arctic-bound waters flowed past an international meeting on controlled ecological life support systems; the brown Samburu in Kenya, where the crocodiles persuaded me to confine my swimming to the game lodge pool.

My raft of life currently bobs at the edge of New Mexico's Gila River. The arroyos, now vigorous with melt, will run only while the snow holds. Their flows carry debris and dissolved minerals down into the Gila. The Gila always flows, with huge pulses after these winter snows, during the spring high-mountain thaw, and after the monsoon storms of midsummer. Last August, monsoon floods on the heels of raging dry-season fires in the peaks upstream turned the river black with liquid ash. Nutrient-rich mud was deposited along and beyond its banks. That mud will nourish riparian plants this spring. A river is not only water. It is particles, minerals, and thus, as the ancient Egyptians knew, a fertilizer of life.

It is tempting to draw an analogy between rivers and blood vessels. Both channel fluids by way of a universal pattern called fractal branching. The air passages of lungs also show this pattern. So do the trellises of trees and the veins of their leaves. Fractals possess self-similarity: Branches sprout small branches that sprout still smaller branches. This self-similarity may contribute to our ability to effectively contemplate life, time, planetary connectedness, and other issues of flow and relationship when gazing into "river" at any of its scales, from the tiniest streamlet to the mightiest river in Siberia.

Let's stay for a moment with the analogy between rivers and blood vessels. The tiny, riffling arroyos around me would be the river's version of capillaries. As these diminutive tributaries carry substances away from the ecosystems around them, so capillaries flush wastes from the cells in their neighborhoods. But the nutrients leached away from the soil ecosystems are by no means "wastes." Such losses are expected, given that the weathering input from rocks cannot build up within the soils indefinitely. The body's blood network also serves to convey a supply of nutrients to the cells. In Gaia the network of branching tubes is employed for only one direction of flow: from mountains to sea. The return flow for the cycle takes place unchanneled in the atmosphere, as winds push around invisible water vapor and giant white barges with cargoes of water droplets. Despite differences in detail, both Gaia and our bodies maintain fluid cycles essential to life at two very different scales.

The similarity between rivers and blood systems should not be overextended. The blood system evolved as a functional part of animal bodies in populations. Gaia is singular; it did not evolve by births and deaths in the Darwinian way. But as we have seen, many of the functions that organisms had to invent for themselves Gaia has obtained for free. To the thermally driven swirling of the atmosphere and ocean we should add the gravity-induced flows of rivers as one of those gifts.

At a landscape scale, rivers serve the ribbons of verdant life along their banks. At a gaian scale, rivers serve the ocean. The mouths of rivers are portals to the bodies of oceans. At these broad mouths the land's losses become the sea's gains. Were it not for rivers and the ions they carry, the oceans would support very little life.

MARINE CYCLING RATIOS

What exactly do the rivers provide for the diatoms, radiolarians, copepods, corals, ostracods, dolphins? Phosphorus had a starring role in life on terra firma. Its birth from rock and subsequent circuits around land plants and soils might be considered the first movement of the biogeochemical symphony written in the key of P. The first movement ends with the whoosh to the sea. There the second movement begins. In listening, we will want to ask whether the marine cycle of phosphorus differs from that on land, and if so, why.

Phosphorus enters the sea in two basic forms. The 1 million tons per year of dissolved phosphate ions is fully ready to be imbibed by blooms of golden-green diatoms or other photosynthesizers. Ten million more tons enters distributed within organic and mineral particles. In a comprehensive inventory of the elements, Elizabeth Berner and Robert Berner, biogeochemists at Yale University, estimate that 2 tons of the particulate phosphorus becomes soluble and is made into additional dissolved ions in the salty waters at the kiss of land and sea. The other 8 million tons of particulate phosphorus is simply too inert to be accessible to marine life. These particulates physically enter the ocean, but before their quick burial they never really enter the ocean ecosystems. As far as marine life is concerned, therefore, 3 million tons a year of biologically available phosphorus enters the ocean.

What determines how much riverine, particulate phosphorus—or nitrogen, or carbon—enters the seas in a form that life can assimilate? And does life have any control over this partitioning, perhaps through an influence on water chemistry? We do not yet have definite answers. But it is certain that rivers deliver the great bulk of the sea's phosphorus. Winds bring some, too—perhaps a tenth as much biologically available phosphorus as the rivers provide. That amount justifies continued scientific study but is small enough for us to subsume it here within the general uncertainty of the riverine inputs.

To compute the marine cycling ratio on the basis of the entry of biologically available phosphorus from rivers, we next need to know the embodiment rate of phosphorus by phytoplankton. Recall that photosynthesis converts about as much carbon into biomass in the sea as on land: about 40 million tons per annum in the sea compared to 60 million on land. The theme now, however, is not carbon but phosphorus. And here the story takes a different turn.

Phytoplankton cells of mixed species can be gathered in fine mesh nets. Sampled from widespread locations across the oceans and analyzed for their constitutive elements, the photoplankton contain many elements that have ratios so constant and predictable that they are considered virtually universal. For example, the ratio of carbon to phosphorus, by weight, in phytoplankton is about 40 to 1. But in the biomass of a typical temperate forest, the ratio is substantially higher, about 320 to 1. Wood bulks large in a forest and is notoriously low in nutrients such as phosphorus. A better approach, then, would be to consider the typical plant requirements for a season of photosynthesis—to analyze what is produced rather than what is standing. This still yields a rather high terrestrial ratio of carbon to phosphorus, about 225 to 1. Yet the percentage of carbon within dry biomass, whether terrestrial or marine, is about the same. Thus the amount of phosphorus is what differs. As an ingredient of total biomass, marine phosphorus is about 1.2 percent, terrestrial phosphorus about 0.22 percent. Although these amounts are both small, their differing by a factor of more than five has major consequences for the cycles.

What causes the difference? Why is so much more phosphorus embodied at sea than on land? This seemingly minor difference is crucial to understanding the operation of Gaia. The difference arises from the functional adaptations of life in water versus those in air and soil.

Terrestrial plants must be structural engineers. Leaves will be whipped by winds, stalks and trunks must counter gravity, roots anchor the plant to the gritty soil—all feats that require husky biological structures. These structures are formed by the truss-like polymers of carbohydrates called cellulose and the tar-like polymers of phenolic alcohols called lignin. Because cellulose and lignin are the number-one and number-two organic substances in plants, and because they consist of carbon, hydrogen, and oxygen, we can see that plant biomass, though poor in other nutrients, is buttressed with carbon frameworks, built tough to stand tall.

The phytoplankton, on the other hand, literally go with the flow—plankton means "wanderer" in ancient Greek. The stresses the tiny green cells confront are not so much physical as chemical. How to maintain a unique chemical interior while surrounded by the salty ocean filled with different concentrations of ions that are either trying to enter or to pull your ions out? Answer: Be a chemical dynamo, a little power pack with a high density of enzymes and molecular complexes that mean action—nitrate reductase, glutamine synthetase, Rubisco. That implies an abundance of nutrients.

Phosphorus is 5.6 times more abundant in marine biomass than in terrestrial biomass. Scaling this number to accommodate the fact that the total biomass photosynthesized at sea is only two-thirds that on land, we can compute the total phosphorus taken up in marine photosynthesis each year, relative to terrestrial photosynthesis. The total phosphorus incorporated in biomass at sea is 3.7 times that on land. In scale-tipping tonnage, terrestrial photosynthesis uses 230 million tons of phosphorus per year. The marine value is thus 850 million tons. In other words, for every man, woman, and child on Earth, 300 pounds of pure phosphorus annually enters the bodies of floating green cells across the seven seas.

Now for the cycling ratio. Marine photosynthesis takes in 850 million tons a year, yet the ocean receives only 3 million tons from rivers. Thus the marine cycling ratio for phosphorus is 850/3, or 280. The biological cycles of the ocean amplify the photosynthesizers by a factor of nearly 300 over what they could do if they used each phosphorus atom only once. This is about 6 times the terrestrial cycling ratio of 46. Impressive.

Why is the marine ratio so strikingly much larger? The marine ratio's numerator is larger: 850 million tons versus 230 million tons. Its denominator is smaller: 3 million tons versus 5 million tons. (Two units of what the land receives as biologically available from rock ends up entering the ocean as part of the inert contingent.) Obviously, both the larger numerator and the smaller denominator contribute to the higher marine ratio. But these are not the reasons we really want. What is it in the dynamics of the marine system that makes the marine cycling ratio so much larger? I will offer an explanation in a moment. But first, it is worth looking once again at how the milkshake overflows. How does the 3 million tons of phosphorus leave the ocean each year?

Unlike the land, the oceans have no rivers to flush suspended phosphorus away. To exit the system, 3 million tons of marine phosphorus must be buried in some solid form each year. According to the Berners, two major inorganic processes emplace phosphorus into the sediments. Some phosphorus gloms onto invisible precipitates of iron oxide in the water. These iron oxides in turn glom onto tiny shells of calcium carbonate, a form of detritus from organisms, while the shells are settling into the sediments. In the second process, dissolved phosphorus is removed in combination with dissolved calcium to form calcium phosphate. This precipitates, falling into the marine muds on the continental shelves and eventually becoming a finely dispersed constituent of shale. These two routes are nonbiological (although it is interesting to contemplate how the output would have to adjust were organisms *not* to drop those tiny shells into the sediments). All told, these nonbiological routes account for about two-thirds of the phosphorus loss from the water. They demonstrate links among phosphorus, calcium, and iron in the biogeochemical symphony.

In addition, about a million tons per year of phosphorus is interred organically, as part of detritus. This organic burial contributes the final third of the annual total. It is worth pausing on this portion. For example, we could compute the cycling ratio by using just this organic component of burial. Out of 850 million tons per year incorporated into marine biomass by photosynthesis, only 1 million tons of organic detritus is lost into the sediments. That gives an "organic only" cycling ratio of 850, which signals incredible efficiency. Should the marine cycling ratio be referenced to the 1 million tons per year lost in organic burial, or to the entire burial of 3 million tons per year, which includes the iron and calcium precipitates? As it turns out, both calculations are useful. The lower ratio of 280 is correct when we are referring to either the total input or the total output of phosphorus across the ocean's borders, and it shows life's enhancement of productivity over the ultimate baseline of phosphorus supply. The "organic only" cycling ratio shows the even higher effectiveness of the ocean-wide ecosystem in recycling its own organic productions.

We are now faced not only with the sixfold increase in the marine cycling ratio over the terrestrial number, but also with the additional factor of three when the marine cycle is referenced to the organic burial only. Looking at how phosphorus leaves the ocean has thus intensified the question: How can marine life in the ocean system so efficiently recapture its vital materials?

Let's return to the same reasoning used for the soil cycle: Life is attracted to other life as sources of energy and matter, because it offers the right proportions of essential elements. Life is thus a magnet for life. This attraction helps keep the cycles nearly closed, weaving life into tightly knit fabrics of exchange. The reasoning explains why the cycling ratios for both soil and ocean systems will be significantly higher than the unamplified, baseline value of one, particularly for a limiting, crucial element such as phosphorus. But the attraction of life for life does not explain why the marine cycling ratio is so remarkably high.

Members of the guild of photosynthesizers are bound to live where the light is. Members of the guild of respirers are attracted to the photosynthesizers. Thus most of marine life lives at or near the surface, including the highest densities of algae, cyanobacteria, and protozoan ciliates and flagellates. Delicate, swimming crustaceans called copepods dart about and feed as relative giants. Bacteria thrive near the surface as well, attached to whatever they can latch onto or just floating free. Through the complex food web of various trophic guilds of respirers, typically ninety percent of the productions of photosynthesis are restored into ionic forms, making the elements available again to photosynthesizers, and within a surface depth of a few hundred feet.

For a variety of reasons, leaks from this surface cornucopia are inevitable. Fundamentally, the water offers no real boundaries, except above at the atmosphere and below at the sediments. Some organisms can use this boundlessness to their advantage. The copepods, for instance, are known to make daily vertical migrations. They rise to feast on the phytoplankton when night cloaks them from predatory fish. Then, during the day, they descend into the safety of darkness well below the photic zone. As they excrete there, they inject organic matter directly into the water far below the surface layer, thus acting as miniature living pumps for directing material downward, out of the reach of photosynthesizers.

Within the surface layers, a variety of materials leak from organisms into the water, some as planned wastes, some as unavoidable losses attributable to life in briny, reactive sea water or to the teeth of predators. When small enough to be called dissolved organic matter, these floating enzymes, sloughed organic molecules, and bits of cells drift as unbounded as the water. Sometimes turbulence mixes these leaked materials down and out of the surface's tight circuits of life. Finally, most prominent of all leaks is the sinking of particles. These include everything that is heavier than water and that is dead and will drop: defunct algae that agglomerate into visible clumps of gooey marine snow, fecal pellets from copepods and fish, even the bodies of dead fish and whales. All these avenues remove embodied energy from the tightly coupled ecosystems at the surface.

What happens to the leaks? Marine life more easily retrieves losses from the ocean's surface than soil life retrieves material lost to rivers. In the ocean, opportunities for life to recapture leaked material are as expansive as the water column itself. Life is there, all the way down. The density of life decreases with depth, as the amount of downward-sifting material itself decreases, having been consumed during its trip into the abyss. As what leaks is consumed, ionic forms of elements are regenerated, primarily by bacteria. But fish are not absent from these depths. For example, the world's most abundant fishes live in these midwaters, species of the tiny bristlemouths. Regenerated in the depths, the leaks are no longer viewed as losses but emerge as parts of the larger cycle. When we expand a volume of consideration downward in the ocean, the computed cycling ratio increases, because the expanding volume includes more material regenerated into the ionic form of any given element. Ultimately, we must include even the creatures at the bottom in the sediments, the final gauntlet of mouths and filters through which not much passes: worms, starfish, ostracods, and bacteria such as *Thioploca*. Once returned to the waters as dissolved ions, the elements are mixed, often slowly but always surely, upward to the surface, thus to re-enter the photosynthesizers and complete the cycle. An essential element may leave the surface in a particle, but at one depth or another, it is very likely to be consumed by life and regenerated as an ion still fully within the gaian matrix of the ocean.

A key to the ocean's high degree of cycling must lie in this pervasive dispersal of life throughout the water column and into the sediments. In the ocean, water is the very habitat. On land, by contrast, water is a component within the gaian matrix of soil. Though necessary for life, soil water is the carrier of leaks. In the ocean, water may remove valuable organic matter from the surface food web, but water cannot take away either the organic particles or the dissolved ions from the totality of marine life distributed throughout the ocean. For marine systems, water *is* the gaian matrix, and life inhabits the entire matrix. The dispersion of marine life is limited only by the flux of nourishment available. The difference in the role of water between the gaian matrixes of soil and ocean would seem to explain the higher degree of cycling that occurs in the ocean.

But if life causes a larger degree of cycling in the ocean, why is the efficiency of productivity in the energy cascade smaller in the ocean? This seems paradoxical. With respect to phosphorus, life is more efficient in the ocean. But with respect to carbon and thus biomass, life is more efficient on land. If the standard of judgment is the cycling of a critical nutrient such as phosphorus, then the ocean takes first place. What if the point, however, is how much biomass can be photosynthesized with how little phosphorus? Then the land wins handily. How should these different perspectives be interpreted?

Let's review the numbers. Each year, the ocean photosynthesizers use 850 million tons of phosphorus to create 40 billion tons of carbon in biomass. Those on land take in 230 million tons of phosphorus to create 60 billion tons of carbon in biomass. Life on land clearly gets more bang for a buck of phosphorus. Even though, as we have seen, there would be greater production of plant biomass if the atmosphere were richer in carbon dioxide, we can still consider carbon dioxide as relatively unlimited and as equally bountiful to all plants from the well-distributed pool in the atmosphere. So it would be a real advantage to use less phosphorus to build the masses of life created each year—masses that, above all, require carbon.

Consider, too, that much of the material photosynthesized each year by land plants is not exactly the most active stuff; cellulose and lignin come to mind—no enzymatic action in those dull molecules. But the structures they build help plants in a big way. Leaves are gracefully and securely stretched out to the sun to optimize the incidence of light upon the molecules doing the grunt work of photosynthesis. A forest or grassland with a leaf area index (total leaf surface compared to a given unit of ground) of 5 or more will absorb 85 percent of the photosynthetically active photons. Putting effort into building the physical scaffolds of stalks, trunks, leaf petioles, and (above all) the cantilevered leaves and needles themselves enables the molecular apparatus of the chlorophyll antennae and reaction centers to be extended as a highly efficient architecture. Thus, shunting carbon into structure boosts the efficiency of molecular systems that actually require phosphorus.

Contrast this situation with that in the ocean. One of the serious limitations of marine photosynthesis is the difficulty of absorbing light. Marine life is hindered by the obstacles it faces as tiny, floating cells in the turbulent ocean. Although most cells have some means of controlling buoyancy, mixing up and down and around is an unavoidable reality of their lives. A lot of light's energy is absorbed by the water itself. In the ocean it is nearly impossible to put out solar collectors at anywhere near terrestrial densities—and thus terrestrial efficiencies. Kelp beds do the trick, but these occur only in some shallow coastal areas, a tiny fraction of the total ocean surface.

Despite their differences in overall efficiency and cycling ratio, both land and ocean ecosystems experience similar limitations or abundances of the essential elements. Elements such as phosphorus that are limiting on land are also limiting in the ocean, because if their flow into terrestrial systems from Vulcan is small, then their flow to the marine systems is even smaller. Elements abundant on land, such as calcium and magnesium, just as abundantly flow into the oceans. Thus, generally, elements that support high cycling ratios on land support high cycling ratios in the ocean, and those with low cycling ratios on land have low ones at sea.

It is also a general rule that the cycling ratio for any given element is higher at sea than on land. This follows from the fact that productivity in terms of incorporated carbon is a little lower in the ocean, but the proportions of nutrients per unit of carbon are much higher. One exception is iron, which phytoplankton spare by forgoing the cellular stores common in plants. But nutrients whose proportions are linked to phosphorus are, on average, between three and four times more abundant in marine than in terrestrial biomass.

These high cycling ratios are a boon to marine life. Take phosphorus. Were the marine cycling the same as on land, marine photosynthesis would be reduced by a factor of six. Much of the ocean is already called a "desert." Were productivity cut to one-sixth, a new word would have to be invented. The high ratios come from living in a big volume of water that can be entirely inhabited. But this turbulent, watery world that facilitates high cycling ratios also constricts life because of difficulties there in absorbing light at anywhere near the efficiencies achieved on leafy land. On the other hand, the land has its own limitation: The life-giving water also leaches nutrients, which then enter the second phase of their lives when they pass to the ocean. Land and ocean thus offer strikingly different opportunities and problems for composing their distinctive versions of the biogeochemical symphony.

A Phosphorus and Nitrogen Duet

Each element has its own song. Take calcium. Calcium ions, abundant in rivers, also enter the ocean from Vulcan's underwater volcanic ridges — an additional input that boosts their overall flux to the ocean by about a third more than that supplied by rivers. Furthermore, calcium is essential not just for the soft tissues of phytoplankton, zooplankton, corals, and fish, but for all white carbonate shells of the shell-making species. When we put the relevant numbers together, the marine cycling ratio for calcium turns out to be about ten, compared to the terrestrial value of slightly more than one.

As on land, the cycling of calcium in the ocean is wasteful compared to that of phosphorus; the ocean is literally flooded with calcium inputs. Yet an imprint of life does appear on the concentration of calcium between surface water and deep water. Dissolved calcium in surface water is about 400 grams (nine-tenths of a pound) per cubic meter of water. In the deep water, say two miles down, calcium is just a tad more abundant, typically 402 grams. This wisp of a difference is created and maintained by what is called the biological pump.

The biological pump consists of a series of steps. First, surface life incorporates an essential element from ions in the sea water. Some of the biological material thus made leaks downward by turbulence or gravity, depending on its size and density. Then, at any and all depths below the surface layer, the biological material is transformed back into dissolved ions. Although the ions return toward the surface via the omnipresent turbulence, the biological pump acts to deplete the surface concentration and augment the deep concentration.

The driving force—what brings an element back to the dissolved, ionic state—is in some cases nonbiological. For example, the tiny cal-

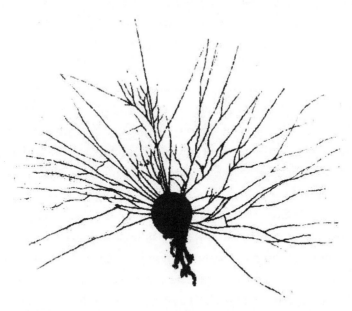

Microscopic web of the sea. This one of many species of single-celled protozoans called foraminiferans was drawn while alive. Its ultrathin, radiating tentacles are used to entangle and thus capture prey, such as floating phytoplankton. Most of the body mass is concentrated within its central, spherical shell of calcium carbonate. Formaniferan shells have formed a substantial portion of carbonate rock during the last several hundred million years, and they have thus been a route for the exit of calcium from the ocean. Actual shell size is that of the period at the end of this sentence.

cium shells of formaniferans and coccolithophorids chemically dissolve back into separate ions of calcium and carbonate only in the very deepest water, where cold temperatures and high pressures make the water undersaturated with respect to the shells. We see the result in the hairline enhancement of the calcium concentration in deep water: about 2 grams per cubic meter of water. In a lifeless world, the calcium concentration would be closer to perfectly uniform with depth. Life has made its presence felt as a faint but definite signature even upon so massively abundant an element as calcium.

The much more dramatic story of phosphorus yields one of the most significant data plots for geophysiology. The regeneration of phosphorus ions in the depths occurs as subsurface deep life feasts on the organic detritus transported downward from surface life. Unlike the dissolution of the calcium shells, this regeneration is biological. But the principal effect is the same: Life depletes ions in surface water and adds them to deep water. Thus we would expect the biological pump to maintain higher phosphorus concentrations in deep waters and lower ones in surface waters.

The numbers have been assiduously measured by chemists on research ships for decades, all over the world and at all depths, because phosphorus (in the form of its ion, phosphate, PO_4^{3-}) has long been recognized as a critical element for understanding marine biology. The maps thereby generated show that life essentially acts as a vacuum cleaner, nearly eliminating phosphorus in the sunlit layer. This is evidence that phosphorus is a limiting nutrient in the ocean, as it is on land.

Deep-water samples are captured in high-tech jugs on strong wire lines, a kind of fishing for water. These jugs can be tripped shut at a chosen depth, from ten feet to several miles below. They are then hauled up and their contents analyzed. The phosphorus in the abyssal waters — say anywhere in the lower half of the world ocean — typically measures about 60 milligrams per cubic meter. That's a mite compared to the mountain of calcium in the same waters. Nonetheless, it is at least ten times the typical surface concentration of phosphorus. To surface diatoms or coccolithophorids, this deep concentration would be the Garden of Eden if only they could get at it.

The jugs that fish for water are also analyzed for many other chemicals. Nitrogen most commonly—and most vitally—occurs as the dissolved ion nitrate (NO_3^-), which also nourishes surface life. Aboard ship, it is analyzed with the same lab equipment as phosphorus. Usually the values of phosphorus and nitrogen are recorded on a foot-wide strip of paper, a chart that alternately shows phosphorus values and nitrogen values, at different depths. The data sheet can also be read as sheet music for the biogeochemical symphony. Nitrogen and phosphorus levels are reminiscent of an ancient Christian chant in which two voices rise and fall in synchrony while maintaining a separation of a given musical interval. Between nitrogen and phosphorus in the ocean, however, the interval is not arithmetic but geometric: a multiple.

At the surface, the mass concentrations of phosphorus and nitrogen are both usually low. As the jugs from deeper and deeper waters are analyzed, phosphorus rises. Nitrogen density can essentially be predicted by multiplying the phosphorus density by a constant factor very close to 7. (In terms of numbers of atoms, or moles, the multiplying factor would be 15.) When phosphorus measures 20 milligrams per cubic meter, nitrogen will weigh in at about 140 milligrams per cubic meter. As a deep-water value for phosphorus is reached, say 60 milligrams, nitrogen will register 420 milligrams per cubic meter. This duet in the concentrations of phosphorus and nitrogen has always been one of the most amazing phenomena to the oceanographers who witness it, as I did in my cross-Atlantic trip. It should thrill anyone who looks at the globally averaged depth profiles of the two elements. For most places in the ocean, take the phosphorus plot, multiply it by 7, and you will very nearly produce the measured nitrogen values. Little else in geophysiology is so predictable.

The reason why the duet is amazing goes beyond the universality of the factor by which their mass concentrations are correlated. As we noted in a previous chapter, the nitrogen-to-phosphorus ratio in the

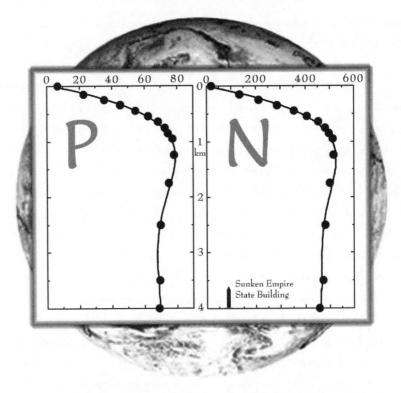

Nutrient ions of phosphorus and nitrogen in the global ocean. Averaged over the Atlantic, Pacific, and Indian Oceans, these plots show global profiles of the two key marine nutrients as functions of depth in kilometers. The values are for milligrams of element per cubic meter of sea water (phosphorus in phosphate ions, nitrogen in nitrate ions). The scale for the nitrogen values is horizontally expanded by a factor of 6.7 from the phosphorus scale to reflect their average proportions in living tissues of plankton. Note the correspondence between the shapes of the profiles when scaled by this living-tissue ratio.

bodies of marine plankton—both phytoplankton and zooplankton, averaged over many species—is also 7. How does it happen that the ratio in the bodies of plankton is virtually identical to that in the water? What is cause and what is effect?

Might the plankton have adapted to build their bodies from what the ocean provides? Perhaps the ocean's chemistry is imprinted on plankton. That answer, however, begs a question about the other elements. Why is calcium, for example, utilized by plankton at a rate only a few times that of phosphorus when it is 5000 times more abundant than phosphorus in the surrounding water? Why do only phosphorus and nitrogen—and not calcium, sulfur, or magnesium—so closely correspond between bodies and water? We must look for other hypotheses.

Might the match between bodies and water be a statistical fluke? Out of a dozen-odd essential elements, some pair within the water could just happen to fall close to their ratio in marine life. Perhaps. But would we really expect the correlation to hold good to a decimal point? Unlikely. In any case, this possibility should be invoked only as a final, throw-up-the-hands retreat, thus only if a reasonable solution is nowhere on the horizon.

Might the plankton themselves have determined the water's composition? By some mysterious yet powerful means, have the plankton written the composition of their bodies onto the chemistry of the ocean in the ink of nitrogen and phosphorus? One suggestion that this should be the working hypothesis comes from examining the ratio of these two elements in land life. We have already seen that the carbon-to-nitrogen and carbon-to-phosphorus ratios are much larger on land. So—what about nitrogen and phosphorus?

A soy plant grown in those pampered hydroponic solutions of the NASA CELSS program has, at harvest, a total nitrogen content about ten times that of phosphorus. That is hearteningly close to the value for plankton, considering the vast difference in size and lifestyle. In a hydroponic wheat plant, the ratio is even better, about nine. What about plants grown in real soils? Analysis of the Hubbard Brook Forest, an experimental ecosystem in New Hampshire, shows a nitrogen-to-phosphorus ratio in the average biomass of about six, which is tantalizingly close to the plankton value.

With the concept of life as a gaian substance promulgated in this book, the ratio would be expected to be relatively universal. For a living metabolism, universal molecules tend to occur in whatever nearly universal proportions are required to run the systems to build more of the universal molecules. Exceptions, of course, are to be expected. Our own bodies, for example, have a N/P ratio of about three. It turns out that the phosphorus content is higher because of the relatively inert storage of structural phosphorus in our bones. But the N/P ratio in our recommended daily food intake is about seven! All told, the point remains: Whether 3, 6, 7, 9, or 10, the ratio falls within a small range; it is remarkably constant. (The ranking is never reversed, for example.) Particularly compelling is the similarity in what is probably the best comparison of all, that between marine plankton and Hubbard Brook Forest, because this compares the natural, whole ecosystems on land with those at sea.

Accordingly, the nitrogen-to-phosphorus mass ratio of seven, more or less, should be taken as a property of life. Thus the working hypothesis should be that life has impressed itself on the ocean's chemistry. But how? By what means?

The answer must include more than the bodies of plankton. The ratios of many elements other than phosphorus and nitrogen in the bodies of plankton are not imposed on the ocean's chemistry. The answer must reflect something special about the fluxes into and out of the ocean. We have already looked at phosphorus. A brief review of nitrogen is in order.

Like phosphorus, nitrogen flows within rivers to the sea in a number of different forms. Again the key is how much is reactive—that is, how much will enter the ocean's chemical system as biologically available nitrogen. We turn once more to the Berners' global inventory, which pegs the number at 19 million tons of riverine, biologically available nitrogen per year. Next, what about the burial flux that removes nitrogen from the ocean? Phosphorus has one biological and two nonbiological burial processes. Nitrogen has only one: Nitrogen is removed as biological detritus, at about 14 million tons per year.

With phosphorus, the river input and burial output were all we needed to know. But nitrogen's many forms and mechanisms for transport among all pools make it perhaps the most complex of all the gaian elements. The main complication for our purposes is that, unlike phosphorus, nitrogen occurs in gaseous forms (primarily N_2). This means that fluxes of nitrogen between atmosphere and ocean must be considered.

Lightning annually converts about 7 million tons of nitrogen from atmospheric nitrogen gas into nitrate that rains into the ocean. Compared to the riverine source, this is not a trivial amount (neither has it been very accurately quantified, but this middle-of-the-road estimate will suffice). There are also a number of gaseous nitrogen compounds — ammonia and various nitrogen oxides — emitted by the terrestrial biota, in particular from soils. These compounds are produced by marine life as well, but they rarely escape from water to the air; instead, they remain dissolved and quickly re-enter the biological cycles. The soil's route to the atmosphere is leakier. Once nitrogen is in the air, the land can no longer claim ownership. Atmospheric gyres waft some of these emissions out over the ocean, where they are also carried down in rain to the tune of 4 million tons per year.

Thus far, the riverine and airborne sources of nitrogen to the ocean total 30 million tons per year, and the sole sink of organic burial is 14 million tons. Although we still need to evaluate one more source and one more sink, it is an appropriate time to pause and take stock of what it is we are looking for.

For the internal amount of a pool to be controlled, either one particular entering flux or one particular exiting flux, at the least, must vary in response to the actual internal amount. For example, in a thermostat-controlled home during winter, whether heat flows from the furnace is a response to internal temperature. Chemical control, too, has been achieved artificially. Sensors attached to the hydroponic solution leaving a bed of growing wheat have been designed to direct nutrients from a stock tank via a tube into the solution automatically, as needed to maintain desired levels of nutrients.

Can any of the nitrogen fluxes we have reviewed thus far directly respond to the nitrogen-to-phosphorus ratio in the ocean? Obviously, the river flux, the lightning flux, and the airborne transport from land gases and particles are not affected by their destination. Nor can I see how the burial rates of either phosphorus or nitrogen could be sensitive to their ratio in the ocean. But two more fluxes of nitrogen between ocean and atmosphere remain to be examined. These are both biological. And both are familiar from previous chapters: denitrification and nitrogen fixation.

One type of habitat for the denitrifiers is where *Thioploca* lives: in the sediments. This bacterium lives at the interface between sediments and water in a region where the water itself is low in oxygen. But everywhere in sediments, below several inches or so, oxygen drops to the anoxic levels required by members of the denitrifying guild. Another type of habitat for denitrifiers is within the water column itself, where bacteria float and consume the organic matter that happens their way, by attaching to it if it is falling, or consuming the dissolved organic molecules. These water column denitrifiers primarily live, again, only where oxygen is low, particularly in places in the Pacific and Indian Oceans. Denitrifiers feed on organic debris and use the oxygen from nitrate to oxidize the debris, a process that locally lowers the deep water's N/P ratio below the standard value of seven. The resulting waste is nitrogen gas, which works its way upward to exit into the atmosphere.

What is the global magnitude of denitrification? Even at the low end of the experts' estimates, 30 million tons of nitrogen per year, denitrification would be the largest single flux of nitrogen we've considered thus far. Furthermore, the upper estimates go much higher. The Berners, for example, propose 70 million tons from the sediment habitat and 60 million tons from the water column habitat, for a total of 130 million tons of nitrogen per year. Before including denitrification, we had accounted for a net positive flux of 16 million tons to the ocean (the difference between the riverine and airborne sources totaling 30 million tons and the detrital sink of 14 million tons). Any of the estimated fluxes of denitrification, from 30 to 130 million tons, would throw the ocean into a nitrogen deficit. The Berners' number, for example, would imply an annual deficit of 114 million tons of nitrogen. With the lower number of 30 million, the deficit would still be 14 million tons per year.

Could denitrification vary in response to the nitrogen and phosphorus ratio? I don't see how. It is usually assumed that this flux will change in response to the amount of oxygen in the ocean. But to the proportions of the two limiting nutrients? Unlikely. This deficit caused by denitrification will prove crucial to the story that keeps marine nitrogen and phosphorus true to their ratio in plankton. But we must first examine the final flux in the marine nitrogen budget.

What remains? Nitrogen fixation, that savior of terrestrial ecosystems, that friend to the farmer—particularly the organic farmer who, for example, alternates corn with legumes. Many planktonic nitrogen fixers such as cyanobacteria float in the ocean. Members of the guild of nitrogen fixers, whether terrestrial or marine, all use the enzyme nitrogenase to convert nitrogen gas into ammonium ions (NH_4^+), which can either be used directly or be converted by other guilds of bacteria into nitrate. Although this process takes place in the water with dissolved gas, the gas would ultimately be replenished with a net flow from the atmosphere into the water.

A key point is that the nitrogen fixers have an advantage only in a situation that would put the nonfixing forms of life under nutrient stress. That would occur when the nitrogen level in the upwelled water is below what is needed to be incorporated at the right proportions with the other elements into the bodies of phytoplankton—in other words, when the nitrogen-to-phosphorus ratio falls below seven. Below the magic ratio, nitrogen would be limiting, and the nitrogen fixers would proliferate.

When the ratio is seven or more, the fixers lose their advantage. Nitrogen fixation is no free lunch on Cafeteria Atmosphere. It costs plenty. There are metabolic costs to make nitrogenase and a host of other specialized enzymes. And the fixers must pay to construct the means to protect themselves from free oxygen, which destroys nitrogenase faster than you can read this paragraph. Members of the nitrogen-fixing guild employ a variety of means to maintain anoxia around their fixation apparatus, including upping their respiration rates to deplete internal oxygen and building extra thick, relatively impermeable walls against diffusion (extra defense costs to keep out the infidel oxygen molecules).

Thus the nitrogen fixers secure an advantage only in conditions that penalize ordinary cells with nitrogen stress. Without such stress on the others, not only do the fixers lack a clear advantage, they are also distinctly disadvantaged (because of the metabolic costs of nitrogen fixation). These dynamics of advantage and disadvantage might exert a thermostat-like control on the ocean's nitrogen-to-phosphorus ratio. When the ratio falls below seven, the nitrogen fixers flourish. They bring a usable form of nitrogen into the biological system—a source that ultimately comes from the atmosphere. When the ratio reaches a value of seven or higher, the nitrogen-fixing activity drops, just as a thermostat shuts down the furnace's production of heat when a home reaches the temperature the homeowner has chosen. The set point in the ocean's biological "chemostat" is none other than the nitrogen-to-phosphorus ratio of the inner guts of life itself, the ratio required by the proportions of little enzymes and other molecules within the universal substance of life.

The story does not entirely belong to the heroics of the nitrogen fixers, however. As we hinted earlier, the denitrifiers play a starring role as well. If the denitrifiers were switched off, then the inputs to the ocean even without the fixers would overwhelm the losses (recall lightning and the other numbers above). The ocean would fill up with more nitrate, and the nitrogen-to-phosphorus ratio would rise above seven. That would be fine for marine life, yet it is not what we see today in the ocean. The denitrifiers are crucial for maintaining the tight ratio because they keep the ocean in a nitrogen deficit, on the average, over long enough time scales. Denitrification chugs along. There will always be plenty of habitats low in oxygen that allow the denitrifiers to thrive by using nitrate to oxidize food in the water and thus release nitrogen gas. In the absence of the fixers but in the presence of the other fluxes, denitrification seems able to de-nutrify the ocean. What the ocean's nitrogen-to-phosphorus ratio was in the distant past no one knows. But these dynamics seem to be the way it all operates today. The dual activities of the denitrifiers and the nitrogen fixers act somewhat like the two types of metal in the spiral winding of a mechanical thermostat; they are a binary system whose elements operate together to ensure that the whole ocean's content of nitrogen and phosphorus stays in line with the proportions in living bodies.

The details of the nitrogen and phosphorus duet, conducted by the guilds of denitrifiers and nitrogen fixers, would be tricky to formulate down to the level of every eighth note. We would need to consider a variety of scales in space. Some regions would be deficient in nitrogen and others not. It is the average, overall dynamics that are important. A variety of time scales enter the composition as well. The ocean could well tip out of balance at any time. This is certainly happening today, with extra inputs of nitrogen via the air as a by-product of fossil fuel combustion and via the rivers as run-off from fertilizer. What we see in the ocean's close proportions of nitrogen and phosphorus is the product of dynamics that probably correct themselves over scales of a few thousand years.

In the nitrogen and phosphorus duet, we witness one of the more colossal phenomena of geophysiology. The ocean is not a planned body, of course; it has not evolved as a whole. Gaia is singular. But within Gaia are evolved organisms doing their biochemical things. The result: a marine nitrogen-to-phosphorus balance that mirrors the composition of life, whole-ocean chemistry determined by the combined effects of two biochemical guilds. Just as the Himalayas are magnificent visible products of the deep dynamics of geology, so the proportional global profiles of nitrogen and phosphorus stunningly portray the underlying dynamics of biochemical guilds.

CONNECTIONS

Playing and listening provide two different experiences of music. In playing, one learns the melodic ups and downs of a single instrument in painstaking detail. My instrument is guitar. I may cop a rock riff from

Eric Clapton or mimic a line from Beethoven. This enables me to appreciate the music's subtleties—for instance, how a well-placed note on the offbeat propels the tune forward, how a later variance of an opening motif unifies the whole, how tension is built and then resolved by the sliding notes away from and toward the key's home tone. Purely as a listener, however, I can also relax and enjoy the complex, indescribable whole.

The same modes of appreciation apply to the global biogeochemical symphony. We can delve into the parts or soar out to the whole. What are the parts? Take your choice. Contemplate the biochemical guilds as the instruments of the grand orchestra of Gaia. Or savor the melodic twists of the elements themselves, cycling again and again with only rare escapes to codas. What is the whole? Gaia itself. But I would not presume to present that entirety. Instead, I will spin out some descriptive sketches, portions of the symphony, a bar here and there drawn from different movements and with only several instruments slated to perform at a time, chosen to provide further insights into the types of chords and polyphonies within the glorious whole.

The duet of nitrogen and phosphorus, a harmony in the realm of Poseidon, expresses a deep and moving concordance between elements. We should look for more duets. One, for example, occurs between sulfur and iron down in the marine sediments. Trios are possible as well, such as the interwoven fates of phosphorus, nitrogen, and iron in various realms of the ocean's surface. That same surface yields its own duet between sulfur and iron. Finally, sulfur can be linked to climate and carbon, in an elegant passage that crosses time scales and bridges space between marine plankton and land plants.

Recall those bacteria called sulfate reducers, in the black, anoxic muck beneath the spaghetti-like *Thioploca* off the coast of Chile. They use oceanic sulfate (SO_4^{2-}) as a source of oxygen to fuel their feeding on the rich organic debris, creating bicarbonate ions and hydrogen sulfide gas as wastes. In so doing, the sulfate-reducing bacteria prevent the burial of a substantial portion of carbon that reaches the sediments. This bio-chemical guild thus profoundly affects the carbon cycle and all else to which it is linked.

In the case of coastal Chile, the waste gas from the sulfate reducers percolates up into the ready membranes of *Thioploca*, who consume hy-drogen sulfide and convert it back to sulfate ions in the water. Some of the hydrogen sulfide gas is grabbed not by bacteria but by abundant, sediment-bound iron. The sulfide gas reacts with the iron to form iron sulfide, or pyrite. Globally, the genesis of pyrite entombs about 40 mil-lion tons of sulfur per year. This is by far the largest exiting flux of sulfur from today's oceans. It is nearly a hundred times greater than the amount of sulfur buried in the coffins of carbon that constitute organic detritus.

How can life participate in shunting such a large flux of sulfur to burial? Does this not violate the rule we have established for the imprint of life on the gaian matrixes—that the fluxes of elements between life and the matrixes will closely follow the embodied proportions of those elements?

In the case of sulfate reduction, life drives a transformation of matter to fulfill a need for energy. Sulfur in this case is not moved into or out of bodies to be assimilated into or decomposed from biomass itself. Rather, the sulfur as sulfate is the oxidizer used by the sulfate-reducing guild of anaerobic bacteria for respiration and thus to power their bodies. Here is the lesson for gaian science: Fluxes of matter tapped by organisms for energy can exceed the fluxes used for material needs. Just when we're nearly lulled asleep by the melodic embodiment rule of material proportions, the timpani of sulfate reduction boom an exception.

Note that without iron, sulfur burial would not occur. Pyrite emerges from the duet of sulfur and iron. Iron is also vital for photo-synthesis, but in an entirely different form from the inert, sedimentary variety that snags sulfur from sulfide gas. It must be in solution. The dynamics of this precious, dissolved form of iron, which have recently been intensely probed by scientists, bring this vital element out of the background of the symphony in certain regions of the ocean. Iron emerges as a soloist.

In surface waters of three vast areas—the southern ocean around Antarctica, the far north Pacific, and the eastern equatorial Pacific— phosphate and nitrate abound. Why don't the phytoplankton take more advantage of the luxury? Do the zooplankton keep the populations of photosynthesizers anomalously in check? Or is another element limiting in these regions? As it turns out, iron is the element in short supply, the limiting factor. Very recently, a once-renegade hypothesis about iron limitation has moved into the hallowed realm of theory.

Clinching empirical support was obtained in May of 1995. A research ship used its propellers like egg beaters to spread minuscule amounts of soluble iron into a Manhattan-size patch of surface water in the eastern equatorial Pacific. Within days the waters turned from clear blue to murky green. Chlorophyll concentrations, a proxy for phytoplankton biomass, increased more than twentyfold. Diatoms in particular proliferated—by a factor of eighty. The numbers of copepods (key herbivores) nearly doubled. The power of the transformation stunned the scientists on board. When the results were interpreted, the scientific community gained a deep respect for windblown dust.

Windblown dust is crucial for the delivery of iron because rivers cannot do the job. Too much of riverine iron is particulate. And the soluble variety quickly slips out of the grip of coastal life into the sediments. Iron riding on dust far out to sea is, to be sure, also primarily inert. But sunlight converts some of it into the soluble, biologically available form as it alights into the photic zones of the oceans.

The three anomalous regions of high nutrients and low productivity seem all to be zones where global wind patterns provide only a meager supply of dust. Comparisons between the delivery rates of iron from wind and from upwelling water still provoke debate about the relative importance of the two sources of iron, but many scientists are beginning to think that the dust hypothesis, if true for the vast southern ocean, could explain Earth's low level of carbon dioxide in the last Ice

Age. An enhanced rate of photosynthesis far into the Southern Hemisphere, primed by iron blown from colder, generally drier and windier continents, may have shifted the carbon cycle into sequestering more carbon dioxide in the ocean, thus leaving less in the atmosphere.

Whatever the ultimate consequences of the iron theory, the recent ocean experiment has demonstrated that a trace metal can indeed be a keystone element of Gaia. This is a case of tiny mass causing mighty effects. The human-induced iron fertilization in the Pacific Ocean yielded one other intriguing result. In the iron-enriched patch of surface waters, the concentration of a sulfur gas, dimethyl sulfide (DMS), more than tripled. This gas came from the biochemical guild of DMS emitters, which responded to the iron boost with a flurry of sulfur notes like a saxophone given a healthy puff.

Dimethyl sulfide is known to have a huge effect on climate. It is released as a precursor molecule into the water from particular groups of phytoplankton. But how much of the gas actually diffuses up into the atmosphere is strongly affected by the complex dynamics of marine ecosystems, including zooplankton feeding habits and bacteria populations. Once in the air, DMS is oxidized fairly rapidly into sulfate aerosol—that is, tiny particles rather than vapor. The aerosol seeds cloud droplets. An increased emission of DMS creates more numerous, smaller, highly reflective droplets. More DMS creates brighter clouds, which reflect more sunlight. The biogenic emission of DMS thus has the awesome consequence of keeping Earth cooler than it would be without the gas.

Consider, then, a world without the biochemical guild of marine DMS emitters. Oceanic clouds would be fewer and more permeable to sunlight. Receiving more solar energy, the oceans would warm. The atmospheric gyres would spread the warmth around to the continents. My calculations indicate an Earth between 15°F and 25°F warmer. Robert Charlson, an atmospheric scientist at the University of Washington who is an expert on the climatic effects of DMS, concludes that an Earth without DMS would be so perturbed from today's state that the resulting climate cannot even be calculated with our current models.

Nevertheless, his best guess is that the temperatures could certainly be boosted even above my numbers.

Conservatively, then, let us use 20°F, the midpoint of the range I calculated, as a working number for the amount of present-day cooling attributable to DMS—and thus the warming that would ensue if DMS were eliminated. A sudden extinction of DMS-emitting plankton would immediately affect clouds and then perturb climate in an interval of decades to several hundred years, thus rapidly producing a world unimaginably different from that of today. But here is where we encounter a crescendo of connections. The relatively simple opening melody from one instrument—DMS—takes on additional meanings as other instruments build on its lead.

In particular, the warming would change the global carbon cycle. This, in turn, would feed back and alter the amount of warming. First of all, a warmer Earth and thus a warmer soil would intensify the rate of chemical weathering of soil minerals into dissolved ions and would thereby lower the carbon dioxide level in the atmosphere by way of marine deposition of carbonates. Second, the lowered carbon dioxide would diminish plant growth, which itself plays a role, independent of temperature, in the weathering rate. As explained earlier, this effect of terrestrial ecosystems is called the biotic enhancement of weathering. The dynamics of these combined processes—biotic and abiotic—can be visualized with the leaky bucket, which we used in Chapter 2 as an analogy for Gaia's inputs and outputs, and in Chapter 3 for the greenhouse effect. This leaky bucket—a bucket with an input faucet, a hole near the bottom for output, and a steady-state water level sensitive to the size of the hole—can be generally applied to the dynamics of a variety of pools in the gaian system. For the situation with DMS we will need not one but two holes in the bucket: one for temperature's effect on weathering, the other for life's effect on weathering.

Consider the bucket's inflow as the volcanic supply of carbon dioxide to the atmosphere-ocean system. The sum of the outflows will be the burial of carbon in the calcium carbonate deposits in the ocean, following the weathering of calcium from silicate minerals in the soil.

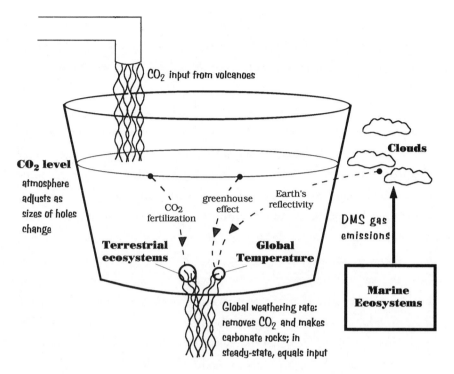

Connections among a sulfur gas (DMS), climate, and the carbon cycle. The leaky bucket provides the conceptual model, here with two holes for the biotic and the abiotic effects on weathering. Factors that influence the size of the holes are shown as dashed lines with arrows to the holes.

What about the two holes? One corresponds to temperature, the other to the biotic enhancement of weathering. Note that the sizes of these holes in our bucket, unlike holes in a physical bucket, are sensitive to the level in the bucket, which here corresponds to the amount of carbon dioxide in the atmosphere-ocean system. Less CO_2 would reduce both the temperature and the biotic enhancement of weathering, shrinking both holes. Independently, temperature would also be affected by the Earth's reflectivity—in other words, by the amount of DMS.

The absence of DMS would raise the temperature and thus enlarge one hole. The carbon dioxide would drop toward a new steady state, thus lowering the temperature and narrowing the hole somewhat. How much? I worked out the calculations of these interactions and arrived at a net warming of only 5°F. When the temperature effect on weathering is included as feedback, the original perturbation of 20°F is reduced by 75 percent to 5°F. But what about the biotic enhancement of weathering?

The lowered carbon dioxide in the bucket, caused by the enlarged hole for the temperature effect, would certainly weaken land photosynthesis. That, in turn, would weaken the intensity of weathering. Why? The answers are several: Life's retrenchment would mean less carbonic acid in the soil as fewer roots and less detritus come to be available for oxidization by soil fungi and microbes; it would also mean fewer ectoenzymes for attacking minerals; and it would mean less humus, which means less water retention—a necessity for the weathering reactions. Thus the second hole would shrink, which would elevate the carbon dioxide and thus restore the temperature back somewhat toward the original value for the warming attributable to the direct climatic effect of eliminating DMS. When I use what I consider reasonable numbers for how the biotic hole responds to the level inside the bucket, the complete system with all feedbacks yields a predicted warming of between 5°F and 20°F, stabilized at about 10°F.

Thus DMS changes the clouds, which change the temperature, which changes the weathering rate, which changes the carbon dioxide, which changes the vegetation, which re-changes the weathering rate, which re-changes the carbon dioxide . . . halt, halt, halt! What does this all mean? And is the effect of DMS 20°F, 5°F, or 10°F? That's a wide range of impacts on climate. I do not have a final answer. But contemplating this scenario shows how the simple question "How much warmer would the Earth be without DMS?" leads to an answer that weaves together climate theory, the carbon cycle, and the role of life in chemical weathering of soil minerals—all in an intricate network of connections within the gaian system.

In music, the meaning of a melody affects what comes along later in the symphony. So too with Earth. The connections between dimethyl sulfide and the carbon cycle are played out over a variety of time scales. How long, for example, would the weathering component take to have full effect? In contrast to the rapid warming that would result from changes in DMS alone, the effect of weathering on the carbon cycle and its feedback on warming would take a hundred thousand years to reach significant strength. But weathering would have its day.

All these separations of the parts of Gaia into smaller units—sulfur and iron; phosphorus, nitrogen, and iron; iron and sulfur again; and sulfur and carbon—are still only bits of the totality, just a few instrumental lines of a symphony. Yet these slices are a way to begin. Paying attention to such details and their web of relations is also the best I can do at this point in my quest to work toward a physiology of Earth. Nevertheless, the exercise reveals the beauty and richness of the music of this sphere. The music beckons. Despite the difficulties we encounter in mimicking it with mathematical models and paraphrasing it verbally, we must continue our effort to hear it. We can always go out into nature and, in the imagination, play some of the better-known bars of the symphony. Then we can sit back, attend to the sky and trees, and know that however rich this experience, the full symphony is a thousand times richer.

8
Gaia in Time

Ten months and seven chapters have passed since my anticipa-
tion of the monsoon that would slake midsummer's drought.
The first green leaves of spring now poke from their protective
buds. The willows are ending their month-long flowering. A week ago
the turkey vultures soared in from Mexico to mock my heaviness with
their aerial acrobatics. Last night, as comet Hale-Bopp blazed in the
west, the first guttural trilling of the canyon tree frogs began.

The level of carbon dioxide measured at Mauna Loa is nearing its
seasonal peak. Galaxies of chlorophyll molecules burgeoning in the nas-
cent green leaves cannot yet compete with the hemisphere-wide respiration emitted by the soil organisms—the deciduous trees from here
to Canada are still basically bare. But within a month, photosynthesis

will dominate. The Mauna Loa curve will start its annual decline. During the next third of a year, atmospheric carbon dioxide will be sucked down. A tiny amount of it will feed the spring and summer grasses and deciduous green unfoldings here in the mountains of New Mexico.

This year's CO_2 peak is bound to surpass last year's, like a high jumper just edging over the bar of the previous record. The trend is up: Each year's peak in May is about one and a half parts per million above the previous peak. In September or October, the annual bottom in carbon dioxide will not be as low as last autumn's. Every three years, this trend results in an atmosphere that contains a percent more carbon dioxide. The upward creep will continue for as long as we furiously burn fossil fuels, keeping the CO_2 graph in contention for the title of most important graph of the twentieth century and probably the century to come. Life—us in this case—is forcing a shift in an atmospheric constituent of Gaia.

A physiology of Earth is intimately concerned with, above all, how the awesome cycles of elements are maintained between life and the gaian matrixes of soil, air, and ocean. Right now the cycles are not in balance. In addition to the upward trend pushed by excess respiration from our fossil fuel energy slaves, the biosphere's breaths are expanding with the decades. This widening of the biotic accordion derives from deeper inhalations, heftier exhalations, or a combination of both. When riding a bicycle this spring, I am confident that the oxygen is at essentially the same level as it was last fall, or the spring before, or, for that matter, on the day I was born. But it isn't quite. Each year the fossil fuel energy slaves consume oxygen. The amount can be not only calculated but measured, though measuring oxygen changes against the vast atmospheric pool is much trickier than measuring changes in the carbon dioxide pool. Other greenhouse gases, methane and nitrous oxide, are rising against their own trace reservoirs even faster than carbon dioxide, perhaps because of agricultural practices and other changes we've wrought on the landscape.

These trends are driven by human activities. Before we came on the scene, all was in balance, no? Nooo. Between 12,000 and 10,000 years ago, as the last Ice Age ended, carbon dioxide surged upward by about a third. That rise was part of a long climatic cycle that restored atmospheric CO_2 against a series of drops in the gas during the prior 100,000 years. Both rises and falls during the Ice Age cycles should be considered as trends, in contrast to the rapid cycles of the breathing of the biosphere within such millennial shifts. Trends—arrows in time—are crucial to understanding the history of Gaia. Trends change the behaviors of the cycles or the levels of the pools: the drop in CO_2 (and thus temperature) since the glory days of the dinosaurs in the Cretaceous, the increase in oxygen about two billion years ago. Taken together, trends such as these reveal Gaia's story.

Let me be frank: it's a murky story at best. Yes, geologists, geochemists, and paleontologists have learned a great deal about Earth history. But debate rages on about some of those ancient trends, and even more about the possible relationships between life and Earth's greatest shifts. The difficulty in *really* knowing what the Earth was like or how it operated a hundred million or a billion years ago is one reason why Gaia's history has not been the focus of this book. I am hesitant to base a theory of Gaia on facts set in the quicksand of the past. There is simply too much solid information to mine from Gaia's here-and-now.

Yet dealing with Gaia in time is essential. Ignoring the temporal element would be like developing a science about humans, bluejays, or sea anemones without a theory of evolution. As I have emphasized repeatedly, Gaia did not evolve. But Gaia contains evolving life. In fact, temporal Gaia is in some ways even more fascinating than the evolution of organisms, because Gaia both contains and is built by the myriad systems of evolution—in other words, the organisms.

What are the fundamentals of Gaia in time? Has biology taken control of the gaian matrixes? Or are we just lucky that life has held together for nearly four billion years? How has Gaia been influenced by Helios and Vulcan, boundary conditions with histories of their own? What is the nature of a system of evolving systems? If the ideas developed about Gaia's here-and-now in the previous chapters are worthy mental tools, they should have numerous applications to Gaia in time.

∽ Gaia's Story

The Archean Age: 3.5 billion years ago

Imagine that a time machine sets you down on a beach beside a lagoon in deep time, 3.5 billion years ago, a few hundred million years after the origin of life. The sediments and rocks of the tide pools are splattered with patches of purple and yellow scum that look like something you'd see at a hot springs in Yellowstone Park. There are even some hints of green among the patches. You've been granted the first look by any human at a portion of earliest life. Better make it a fast look. You'll soon be writhing on the ground, gagging, dying!

It's not the heat that would get you, or the humidity, though you would not want to linger long in the wet sauna of Earth's early Archean age. How hot was it? Welcome to big mystery number one in the history of Gaia. Much more carbon dioxide was almost certainly in the atmosphere, making for a massive greenhouse effect. Imagine an average Earth surface of 100°F to 120°F (about 40°C to 50°C), compared to today's pleasant 60°F (about 15°C). The world before all life could have been even hotter, 150°F (about 65°C) or more, because microbial crusts on land would have already exacted some cooling by their enhancement of rock weathering.

The air would make you writhe with your first breath. Your gasping would be due to a problem (for you) that virtually all historians of early Earth agree on: There's no oxygen, or very little. Exactly how little is a subject of cantankerous debate. Whether oxygen was at one-hundredth or one-millionth of today's level makes a vital difference in the constraints scientists can place on the chemistries of air, water, and rock. But your time-traveling self is a goner no matter which degree of anoxia prevails. A methane smog may act as a shield against ultraviolet radiation in the absence of ozone (made from oxygen). Whatever the Archean air may be, it is certainly unlike any we have ever breathed. Nevertheless, the colored mats of microbes at your feet are carrying on just fine.

We know that the archean Earth spun faster and once had days only fourteen hours long. Also, the moon was closer and thus looked more like a pie plate in the sky than a quarter. But the onset of basic biology is far harder to reconstruct. Did life come from space packaged in the pores of meteorites? Did life first stir deep within the Earth — remnants may still lurk there — and percolate up to the surface later? Did life form upon clay sediments in that lagoon?

Wherever life arose, geologist Euan Nisbet, of the University of London, makes a case for hydrothermal vents as the sites of the earliest complex biomes. The evolution of key enzymes, he maintains, occurred at these hot microbial gardens of Eden. Inhabitants were strictly prokaryotes, of course, ancestors of both bacteria and archaea. Imagine physical settings along deep ocean ridges, like the stunning and exotic vent ecologies discovered in recent years. But also imagine prime habitats along the percolating flanks of broad-based shield volcanoes, such as those of Hawaii, which offered a variety of water depths, including shallow but still potent hydrothermal flows. What makes a variety of vent sites so attractive as homes for the earliest life, before the arrival of photosynthesis and biologically controlled, global cycles of nutrients, is that they were the richest sources of chemical energy and crucial elements.

Nisbet and others point out how the history of evolution at hot, underwater vents might be written on today's genomes and in the catalytic cores and assembly strategies of the little enzymes that run the world. Elements in enzymes that handle nitrogen include iron, sulfur, manganese, copper, and magnesium — all much more reliably abundant at vents than elsewhere. Chlorophyll may have its ancestry in a heat-sensing system. An ancient chlorophyll similar to today's bacteriochlorophylls, which absorb in the infrared, would have been useful in helping motile microbes position themselves just so within the intense and shifting thermal gradients. And why is it that proteins related to those that repair cellular mechanisms in the wake of disruptive heat shocks are ubiquitous players in a number of diverse functions in today's cells? It is because at life's origin, such heat-shock proteins may have been essential equipment.

Evolutionary modifications could have turned the ancestral heat-sensing chlorophyll into chlorophyll-a, which absorbs visible, higher-energy photons. Making use of the higher energy gifts of the sun, these bacterial innovations were able to split water, thus obtaining both a source of hydrogen feedstock and ion gradients for energy storage across cell membranes. Many scientists suspect that such oxygen-generating photosynthesis could have developed very early, liberating life from dependence on chemosynthesis at the vents. Green cyanobacteria, or something that looks much like them in the ancient fossil record, absent the color, would have lived not at the vents but in the shallow waters of lagoons and bays. They also would have floated across the open seas wherever nutrients provided sustenance.

The green light-lovers would have harbored the molecular match-maker of photosynthesis, Rubisco, today helped in its assembly by a molecule called a chaperonin, a relative of the heat-shock proteins. This suggests an ancient vent heritage for what was to become the cardinal enzyme of the biosphere. As a complex protein, Rubisco has evolved and diversified across time and species. Yet as a functional molecular complex, it carries a hydrothermal fingerprint in its iron-sulfur center, as well as in the molecules from the heat-shock tribe that help in its assembly. Rubisco thus demonstrates a continuity in the deep dynamics of Gaia that extends back nearly four billion years. So does chlorophyll. As a simple pigment, chlorophyll has probably been faithfully conserved in structure—a heritage that may very well date from soon after the origin of life. Look at a leaf and gaze into the living presence of deep time.

The biome communities of vents, shallow waters, and open waters may have assisted one another. Chemosynthetic vent communities, for example, would have been sources of several elements essential for photosynthetic surface biomes. These elements traveled by the ocean's mixing of bacteria away from the vent communities. Nisbet suggests that bacteria themselves would have been care packages of iron for the nutrient-starved surface communities. Those surface dwellers, in turn, through photosynthesis, produced oxygen. Early on, the oxygen would have been captured by purely chemical processes, such as oxidizing hydrogen sulfide and other substances, thereby limiting oxygen to a trace gas. But the sulfate ions so produced, mixed downward by the ocean swirls, would have been a valuable source of oxidizer to the microbes at the vents. Gaia's free circulatory system probably played an essential role in early evolution.

What about life on land? It is possible that continents were relatively tiny, and more of the land was submerged. Imagine an Earth of numerous Japans, Englands, and Indonesias. With so much of the land's area near warm surface water, rainfall would have been bountiful. Some pioneering bacterium blown up on land from the ocean surely would have taken hold on the rocks. Today such microbial terrestrial communities inhabit the deserts in places where plants cannot live. These "cryptogamic" microbial crusts are essential agents that hold thin soils against erosion. In a world with no competition from plants, the cryptogamic communities could have been widespread. If so, they were probably the first biotic enhancers of weathering. Geochemist David Schwartzman and I have concluded that the surface temperature of Earth in the earliest days of land life, though hot by our standards, was nevertheless significantly cooler than the hell it would have been without this microbial crust.

The rise of oxygen: 2 billion years ago

Did the full orchestra play from the first few dips of the baton of life's creation? In other words, were all the biochemical guilds in existence by the time life was able to range from ocean vents to lagoons and on to land? NASA scientists Rocco Mancinelli and Chris McKay think that one important guild in the nitrogen cycle did not come into being until oxygen had risen to some critical level, perhaps two billion years ago. Not until there was enough free oxygen to make the biota much more efficient and thus productive did a crisis ensue in the supply of biologically available nitrogen, a crisis that would have impelled the evolution of nitrogen fixers. Nisbet, however, believes that the crisis that ushered in nitrogen fixation began much closer to life's origin.

Whatever the time sequence—and much challenging research lies ahead in figuring out the steps of biochemical evolution—the principle would probably be: starvation breeds innovation. Buckminster Fuller's apt slogan for human technological progress could as well be applied to metabolic evolution: emergence by emergency.

In addition to triggering a possible nutrient crisis, the rise of oxygen opened a tremendous opportunity for increased productivity and thus organismic evolution. Dick Holland, a geochemist at Harvard University, has studied a number of changes in the chemistry of fossil soils, ores, and sediments formed in ancient days. He has concluded that be-

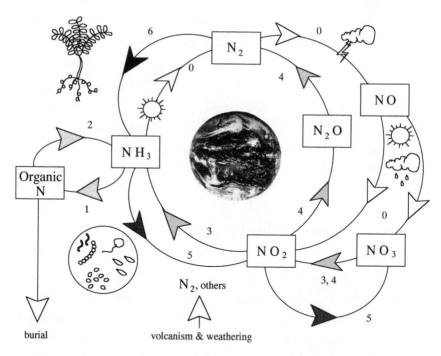

Evolution of the nitrogen cycle. The sequence of stages begins with (0) prebiotic lightning, solar-induced changes, and raining out of ions. The remaining stages are all evolutionary innovations by life: (1) ammonia assimilation, (2) ammoniafication, (3) nitrate assimilation, (4) denitrification, (5) nitrification, (6) nitrogen fixation. This sequence of biochemical guilds follows the theory of Mancinelli and McKay. Theorists debate the details of the sequence, but they all seek to explain innovations as having logically followed conditions created by previous innovations, either within the nitrogen cycle or through connections to cycles of other elements.

tween 2.2 and 1.9 billion years ago, Earth's atmosphere changed dramatically. Oxygen rose from a trace to perhaps around a sixth of its current level, enough to shift life into an aerobic state. Fossil evidence and computer-generated histories of genetic splits indicate that the aerobic eukaryotes—diatoms and amoebae are living examples—evolved at this time.

What caused the rise in oxygen? David Des Marais of NASA's Ames Research Center posits that a surge in mountain building fostered a rush of sediments, which in turn upped the rate of carbon burial offshore. James Kasting of the Earth System Science Center at Penn State University holds that as oxygen-sponging gases from volcanoes fell below a threshold, the level of the now-vital gas skyrocketed. Operating alongside these factors may have been positive biological feedback: More oxygen spawned a more productive biota and thus more burial of organic carbon, which left behind more oxygen. The elevated oxygen level, once reached, would have sustained enough life to create and bury the carbon to maintain those elevated levels of atmospheric oxygen—a sort of self-stabilizing attractor state in the dynamics of Gaia. Whatever the answer, it appears that oxygen never dropped back to its previous level as a trace biogenic gas. At this watershed in Gaia's story, oxygen became a major and enduring constituent of the atmosphere.

A second oxygen pulse: 1 billion years ago

Oxygen surged upward in more than one step. Donald Canfield and Andreas Teske, researchers at the Max Planck Institute for Marine Microbiology, have found evidence that a second pulse occurred sometime between a billion and 600 million years ago, during another interval of increased carbon burial. By examining sulfur isotopes in ancient sediments, Canfield and Teske have concluded that before this pulse, not enough oxygen penetrated the ocean for sediment microbes to oxidize the hydrogen sulfide gas emitted there by others. In surface waters lived photosynthetic sulfide oxidizers, but no nonphotosynthetic sulfide oxidizers (such as *Thioploca*) occupied the sediments. Indeed, Canfield and Teske's map of genetic branchings shows that *Thioploca* and its sibling bacteria genera diverged precisely around the time of the late Protero-zoic oxygen pulse. Apparently the biochemical guild of nonphotosyn-thetic sulfide oxidizers began at this time, a new instrument in the bio-geochemical symphony.

Did the nonphotosynthetic sulfide oxidizers actually come into ex-istence at the second oxygen pulse, or did they merely diversify and proliferate then, after existing for perhaps billions of years in small pock-ets of the environment? This is a fascinating issue. But in either case, the oxygen pulse seems to have allowed their flowering into a large-scale presence in the planetary cycles. Most guilds may have evolved very early. The relevance of all guilds in Gaia's story becomes clear when each plays a part in the planetary dynamic. Piecing together the stages in the evolution and proliferation of the biochemical guilds that arise (and sometimes decline) is the core research needed to fill in the many pages missing from Gaia's story. Delineating how these geophys-iological capacities sequentially arose will offer fruitful new views on the evolution of complexity in the biosphere.

What other guilds have added sequentially to the tale? At the end of the Proterozoic, about 600 million years ago, multicellular (metazoan) life evolved, probably facilitated by the higher oxygen levels. Perhaps a cooler planet, as well, fostered this biological step. According to David Schwartzman, cooling to a threshold comfortable for metazoa was a precondition for this more complex form of life. We might be tempted to dismiss the metazoa as just the same old biochemical stuff—aerobic bacteria and archaea writ large. The metazoa could do only what mi-crobes had been doing for billions of years. Indeed, the performances of metazoa seem downright shabby, compared to the biochemical py-rotechnics of bacteria and archaea. No plant, fungus, or animal can, for example, fix nitrogen or generate methane, though some of us provide lodgings for bacteria that possess these talents.

Rooted photosynthesizers: 400 million years ago

Mark and Dianna McMenamin, paleontologists at Mount Holyoke Col-lege, see fungi as fundamental to the emergence and evolution of macroscopic land life. They declare that land life works as an integrated system of fluid exchange, which they have christened Hypersea. Because land creatures lacked the ocean's generous bath of ions, they had to access and store fluids, and transfer them internally. Fungal threads are masters of such Hypersea functions. They connect the roots of their symbiotic plant partners, generating vast underground networks of nutrient channels that often link disparate species. With the emergence of Hypersea, plants and fungi launched an intricate partnership that has lasted 400 million years.

One of the rounds of the dance included the invention of structural materials capable of raising a redwood tree. The toughest of these is lignin in wood. Data assembled by Earth system scientist Jennifer Robinson indicate that trees became weighty producers of lignin long before other organisms developed truly effective ways of degrading it. Today, for example, certain fungi (predominantly the white-rot basidiomycetes, which include fungi that make mushrooms) have enzymes in their molecular locksmith bags that can crack open lignin. In the interval—perhaps 100 million years—between the invention of lignin by plants and the evolutionary scaling-up of lignin-cracking enzymes, massive amounts of carbon were buried. This lag allowed oxygen levels to climb to peaks never again reached, perhaps fifty percent higher than today. High levels could have fueled the bodies of giant dragonflies—with two-foot wingspans—during the Carboniferous, an apt name for an era rich in fossil carbon.

The evolution of land plants apparently changed conditions for life in coastal seas as well. As documented by a team led by Thomas Algeo of the University of Cincinnati, in the short interval from 400 to 380 million years ago, the size of the biggest land plants soared from inch-short dwarfs to hundred-foot giants, an amazing case of evolution in action. At just about the time trees were reaching skyscraper status, an unusual series of marine events took place. Today we read the evidence recorded in sediments of many shallow-water sites: episodes of black shales. Black shales usually indicate anoxic conditions in the water. Did a continental flux of nutrients from tree-enhanced chemical weathering create permanent algal blooms along the shorelines that overwhelmed the water's oxygen?

Size alone does not transform an evolutionary invention into a biochemical guild of global consequence. A guild must appreciably influence the gaian brews of soil, atmosphere, and ocean. Consider the trees and land life in general. More, consider their symbiotic fungi, perhaps the *sine qua non* for the vascular plants and thus ultimately deserving credit for whatever the trees can do. As we saw from the leaky-bucket model of carbon dioxide dynamics, an increase in the size of the hole (which represents the effect of terrestrial ecosystems on the intensity of weathering) yields a lower level of fluid (carbon dioxide) in the bucket. In thicker, humic soils, mineral grains are coated with water for longer periods of time, yielding fuller weathering. Solid support for increased weathering caused by vegetation is coming from experiments at Hubbard Brook Experimental Forest in New Hampshire. We usually think of trees as defined by trunks, branches, and leaves. But as a recent biochemical guild of Gaia, it may be tree roots that are most significant in altering the cycles of elements. In the largest taxonomy of the guilds, trees are photosynthesizers, big siblings to algae and cyanobacteria. But they occupy a unique slot as *rooted* photosynthesizers. Rooted photosynthesizers have worked distinctive effects on the cycles of carbon and other elements.

Gaia's story weaves together changes in life and the gaian matrixes. Environmental conditions change over time, producing new opportunities for guilds to evolve and expand, including significant variants on the old guilds. In this version of the story, biochemical guilds are sequentially added to Gaia.

Biochemical guilds are most simply defined by what their members produce or degrade. But as we have seen with land plants, the telling effects of a new guild can also derive from how their very form alters the enveloping chemical brews. Another fairly recent guild, the one with which I will end this story, is distinguished by *where* its members put their wares.

Planktonic calcifiers—calcareous plankton—enter the story in droves only during the last several hundred million years. They include the coccolithophorids, which are photosynthesizing algae, and the planktonic formaniferans, the microscopic spider webs of the sea that capture prey with their sticky, radiating filaments. All are single-celled. What makes these diverse creatures a guild is that they precipitate calcium carbonate as shells around their tiny bodies. Moreover, they float at the ocean's surface. They thus know no boundaries and wander with the currents.

To precipitate carbonate, and yet to float, was a combinatorial innovation that came late in Gaia's story. Before the proliferation of the calcareous plankton, calcium carbonate was buried almost exclusively in nearshore environments. The creatures that were so buried included the microbes that made stromatolites in ancient times, reef-building algae, sediment-dwelling formaniferans, and coral. Recall that carbonate burial is the main route for removal of carbon dioxide on geological time scales, surpassing the burial of organic carbon. Exactly where sediments of calcium carbonate accumulate, therefore, makes a great deal of difference in how rapidly and effectively CO_2 is recycled to the atmosphere through volcanism. Many carbonates deposited eons ago in shallow water—on continental shelves or inland seas—have never been recycled to carbon dioxide. They may well be uplifted and probably dissolved into ions in rivers, carried back to the sea, and simply reburied as calcium carbonate. Very rarely will they be pressure-cooked enough to release the rock-bound carbon dioxide back to its gaseous state.

On the other hand, when calcareous plankton die, their shells drift downward from anywhere on the ocean's surface. Bottoming out where the ocean floor rises above the deep zone of dissolution (for example, on the tops of the wide areas of ocean ridges), the shells become parts of the ocean's tectonic plates. Frosted with mile-thick calcium carbonate sediments, the plates slowly move toward subduction zones. There Vulcan provides Gaia a gift of hot and pressurized ruminant stomachs. Some fraction of the carbonate plunges deep into Vulcan's interior. But a remaining fraction of the subducted carbonate—a fraction much larger than that possible with shallow burial—is regenerated into carbon dioxide, which percolates up by coastal and island arc volcanism. My own work has shown that the evolution of calcareous plankton could have boosted the overall flux of carbon dioxide generated by volcanism. Without the calcareous plankton, therefore, the world might today be substantially colder.

That's Gaia's story in a nutshell, as I see it. The story is complex and for now can only be sketched. Firm facts must be blended with deft interpretations. Yet surely life would have both depended on and influenced the gaian matrixes from the very beginning. Future versions of the tale will tell us more about all the biochemical guilds and will reveal subplots detailing the dynamics of essential elements. For example, how much have the largest cycling ratios changed over time? What crises and opportunities arose during the nearly four billion years of life's history? We can count on much illumination and many surprises in the years to come.

THE CONTINUITY OF GAIA

The complexity of the biosphere is built on excrement.

—Euan Nisbet

It is one thing to tell stories about nature. As winter turns to spring, paragraphs and poems could be written about the emerging leaves of different species: willow's sheenless, oar-like blades; cottonwood's sticky, sweet-smelling unfurlings; box elder's notched tripartite banners. Such detailed descriptions, however evocative, would leave scientific curiosity unsatisfied. Tellers of science tales seek unifying principles. They might point out, for example, how the papery green sheets emerge from compact, protective buds. They might highlight the geometric course toward maximizing surface area for capturing sunlight.

The same goes for Gaia's story. It is one thing to recite a chronicle of events. It is quite another to root out underlying trends and discover unifying principles. A central feature of Gaia to ponder is the unbroken line of life for nearly four billion years. What principles have governed life's tenacity? What dynamics, if any, prepared the way for such persistence? In addition to having ferreted out many tentative details in the story, can we frame generalizations that will abide in the face of future refinements—and even overturnings—of our understanding of the details?

What about Gaia's complexity? Surely it integrates the evolution of organisms. A tale about complexity would include the rise in C_4 photosynthesizers with the concomitant drop in carbon dioxide, and the evolution of sulfide-oxidizing bacteria spurred by escalating levels of ocean oxygen. But what can be said that is deeply general about the intricate connections among life forms and the gaian matrixes across the vast plains of time?

Before the origin of life, the goddess Fortuna touched this planet. If it had been too big, Earth would have resolved as a gas giant with a hydrogen atmosphere. Too small, and it would have petered out volcanically like Mars. Because it is geologically inactive, Mars cannot recycle carbon dioxide from carbonates. Mars is cold for another reason as well: It is too far from the sun. An Earth too close to the sun would have suffered another type of disaster, not only from the extra solar gain but also from speedy loss of water as vapor was split in the stratosphere and hydrogen escaped. Hot Venus is bone dry. Planetary scientists allude to the Goldilocks fable: Mars is too cold, Venus is too hot, and Earth is just right.

But what defines "just right"? Earth was warmer in the past and was almost certainly very much warmer in the distant past. "Just right" is a temperature range—basically anything suitable for surface life, even if that life is limited to thermophilic microbes. So what makes Earth the temperature it is today? For that we need to look at how Helios, Vulcan, and life within Gaia have changed through time.

According to current astrophysical models, the sun's output has grown by 30 percent over the past four billion years. By itself, this increase spurred a warming trend on Earth. But Vulcan also played a role in the game of long-term climate change. As Earth's interior cooled over time, the release of gases lessened. As we noted earlier, a gradual decline in oxygen-hungry gases such as hydrogen sulfide abetted the rise of oxygen. Vulcan's carbon dioxide emissions belched from the depths would have gradually dropped as well, cooling climate. Vulcan's story affects Gaia's in a second way. Continental areas, as accumulated slag heaps from plate tectonics, expanded over time, which increased the area available for rock weathering, thus lowering carbon dioxide. By two means, then, Vulcan's history contributed to cooling.

Which trend dominates, the warming from Helios or the cooling from Vulcan? To examine this question, we need a model for the carbon cycle similar to the familiar leaky bucket but coupled to a formulation for how the sun affects climate. David Schwartzman and I have assembled the relevant model. Here is a plausible scenario I have run. Start with a temperature at 3.8 billion years ago of about 150°F (65°C). Keep everything constant except for the model sun, whose energy output gradually increases by 30 percent. Today's temperature would then be 160°F. Now restart at the initial conditions, and add Vulcan's two cooling trends. Schwartzman's best estimate expands the continents by a factor of four and drops volcanism threefold—overall a twelvefold impact (both trends diminish carbon dioxide). From this number, we might suspect that Vulcan dominates the combined system. But numerical output requires an explicit model, because the sun affects the energy input directly, whereas the deep Earth affects the greenhouse effect. Nonetheless, under the influence of both Helios and Vulcan, the initial temperature of 150°F drops to 120°F (50°C). The diagram shows the full histories. Vulcan's cooling does overcome Helios's warming to some extent.

Today's average global temperature is a cool 60°F (15°C). I could have set the initial temperature at a lower value, one that would have made the trend over 3.8 billion years reach 60°F like an arrow dropping into a bull's eye. But then no room would have remained for the third and probably most powerful factor in Earth's temperature history: life. Land life enhances the chemical weathering of rocks. The liberated calcium ions from silicate rocks entomb carbon dioxide through the burial of calcium carbonate at sea. Best estimates currently put the boost to

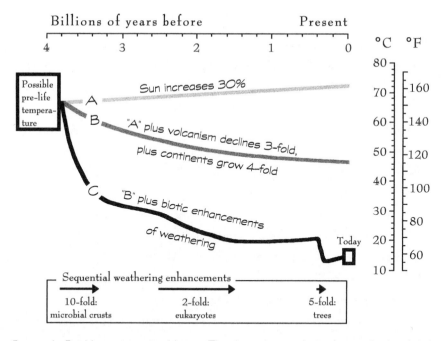

Factors in Earth's temperature history. The three curves show changes in the global temperature over nearly 4 billion years for three scenarios. Scenarios A, B, and C may be thought of roughly as: Helios alone, Helios plus Vulcan, and Helios plus Vulcan plus Gaia. Progressive increases in the biotic enhancement of weathering, here corresponding to a series of evolutionary innovations, were essential to cooling the planet to the current average surface temperature of about 60°F (15°C). The overall enhancement of weathering by life over the abiotic rate is 10 × 2 × 5 = 100.

weathering that accompanied the evolution of deeply rooted terrestrial ecosystems—trees—at a factor of seven (with a probable range from three to ten). If seven is the number, then this would have had roughly the same effect on carbon dioxide and thus temperature as Vulcan had over time.

But the biotic enhancement with today's verdant land life compared to no life at all is probably even larger. Microbes alone can stabilize thin soils. Even a veneer of soil, of small grains with large surface areas, lengthens the times of exposure to water, a necessary condition for weathering. Furthermore, microbes release enzymes that mine minerals. The crucial number is the overall difference between a world with no life—mostly bare rock—and the deeply rooted soils of today.

Schwartzman and I place the number at a factor of a hundred or more. For the resulting effect on climate, I will partition the enhancement across a series of steps: tenfold during the first half-billion years after life's origin, with the formation of the earliest microbial crusts in the hot sauna world; twofold more at the evolution of eukaryotic cells around two billion years ago, which facilitated terrestrial productivity; and five-fold more during the evolution of upright vascular plants around 400 million years ago. Ten times two times five — that's one hundred overall.

The results are shown on the graph. To review: All scenarios start with the same pre-life temperature of 150°F. The warming sun alone makes the computed "present" temperature 160°F. Adding the two cooling trends of Vulcan brings the "present" down to 120°F. Finally, adding life lowers the "present" still further to today's actual value of 60°F. Any way you slice it, life is a major player. Were the overall enhancement to be as low as ten, then life's effect would be roughly equal to the impact of Vulcan. More probably, life is the coolest player in the game.

This section began with a promise of generalizations about life and Gaia. What can be culled from this temperature history?

All of life is affected by Earth's temperature. The temperature of 120°F is the upper limit for the most heat-resistant animals and plants. Without the biotic enhancement of weathering delivered by ancient microbes, creatures such as butterflies and maple trees might never have evolved. A high temperature would have precluded the kinds of enzyme systems that are deployed, and perhaps required, by the larger organisms. By cooling the planet, early life paved the way for later life.

Note that the biotic enhancement of weathering would not have been a one-time invention. It is, rather, a general strategy for accessing minerals and a happy by-product of life's more definite intentions. Consider: liberating nutrients from minerals is beneficial to the organisms that accomplish it. Any organism — bacterium, fungus, or plant — capable of secreting an acid that extracts nutrients from rock will survive and propagate. Promoting a deeper soil is also advantageous for life. Thus the biotic enhancement of weathering would inevitably have been encouraged by the evolutionary process, operating on the scale familiar to Darwin—that is, at the level of organisms, not planets. Enhancements probably ratcheted upward in stages. As a side effect of this push, the hole in the leaky bucket of the atmosphere-ocean's carbon dioxide system was progressively widened, which progressively lowered the steady-state level of carbon dioxide and progressively cooled climate.

The terrestrial organisms do not, of course, know they are cooling the planet. Nor do they care. They only "know" that ions are easier to come by this way. The adaptive property for which the organisms are selected is very different from the big climatic consequence of that selection. Evolution for nutrient gathering—however accomplished, by secreting acids or facilitating deeper soil—had the powerful side effect of cooling the planet.

Finally, we have come to what may be a foundational principle of Gaia in time: What organisms do to help themselves survive may affect the planet in enormous ways that are not at all the reasons those survival skills were favored by evolution. For the organisms, the survival strategies virtually always carry a metabolic cost. The gaian side effects come for free. This direction of thinking promises a sublime insight far beyond an analogy between Gaia and an organism. The makings of a physiology and a geophysiology are very different. Yet the two evince ties that bind. The strong ties between parts of our bodies, honed by millions (or billions) of years of evolution, need hardly be mentioned. The ties between parts of Gaia can be just as strong, because life is chemically connected by the whirls of air and water, by the porosity and bioturbation of soil. When land life cools the planet by a new advance in weathering enhancement, all life in the ocean is affected. The connections within Gaia are intimate and deep. If Schwartzman and I are right, then major features of the biosphere—such as today's relatively cool temperatures and the existence of plants, animals, and fungi—are due to the free effects on the carbon cycle from life's evolved quest for soil minerals. The global effects of gaian developments are no-cost by-products of local, self-centered, organismic evolution.

For a second example of the relationship between gaian developments and selection at the level of organisms, we can turn to the emission of dimethyl sulfide (DMS) by phytoplankton. Ken Caldeira, a climate researcher at Lawrence Livermore National Laboratories in California, was curious about whether plankton could be selected to produce DMS specifically because of its climatic effects. Recall from the previous chapter that this biogenic gas, after being released into the water as precursor molecules from the bodies of *some* species of plankton, disperses into the atmosphere and, transformed into sulfate aerosol, increases the reflectivity of clouds. Could the more reflective clouds benefit the plankton underneath? Perhaps more nutrients are stirred up by the resulting cooler water. Caldeira wanted to weigh these presumed benefits against the metabolic costs to the algae of producing the precursor molecules. He liberally estimated an enhancement of growth from an increased stirring of nutrients. He conservatively tallied the metabolic costs. The finding: no contest. Metabolic costs outweigh climatic benefits by a factor of a billion or more. With such a skewed ratio, cheaters would proliferate. They could live mixed in with the DMS producers and derive all the benefits of upwelled nutrients without paying the huge metabolic costs. The lesson: Phytoplankton must synthesize the DMS precursor solely because it benefits their individual growth while it is inside their bodies, taking no consideration of the complicated and diluted route of influence through the atmosphere. The precursor has indeed been proved to help cells regulate their ion contents relative to the surrounding salty water. Plankton don't even want to release DMS; it is forced from them in predation by zooplankton or bacteria. The survival-promoting, internal function of the DMS precursor is why the genetic heritage of synthesizing it is passed on by the generations, not because it has a climatic effect as a gas spreading across the sky.

What if DMS were found to be detrimental to marine life? More reflective clouds, for example, dampen photosynthetic potential by reducing the light that reaches the surface. In this case, the DMS emitters are actually hurting all the other life in their locale. But the emitters would still keep on emitting because of the huge survival benefits of regulating their internal ions. The numbers are of the same magnitude as before: The climatic detriment would be only a ripple on the ocean of the real evolutionary math going on within the organism. On the gaian scale, whether DMS as a diffuse gas causes large-scale benefits or detriments (or both) may not matter, because in any case the climatic effects forge intimate links among all organisms living within the DMS-determined climate. A world that is cooler because of DMS would have different climate zones, rainfall patterns, and ocean circulation. If the world average temperature is now 10°F cooler because of DMS, and 60°F cooler because of the progressive biotic enhancement of weathering, then the whole living world is to some extent adapted to a physical reality vastly influenced by some of life. All of the tens of millions of species are united by DMS and the biotic enhancement of weathering. The situation is awe-inspiring: Neither biogenic DMS nor a biotic enhancement of weathering evolved *because* they cooled climate, and yet their existence perpetrated free gaian effects that profoundly link all life. Gaia is (probably) built from free by-products, side effects.

The ultimate in free by-product is excrement. Buffalo dung, to be sure, participates in a feedback loop that partially supports the buffalo: nutrients are provided to grasses and thereby returned to grazer. But buffalo do not defecate to maintain this loop or to feed dung beetles or fungi. They defecate to rid themselves of useless, poisonous, and burgeoning wastes. If the wastes were useful to it, the animal would reuse them directly, not portion out the benefits to other organisms such as cattle or deer via grasses. (Rabbits, in fact, do re-ingest their own droppings, because their hind guts can extract some nutritional benefit from a second round of processing.)

For a science of Gaia, excrement must be taken beyond its usual sense to include all that is excreted, meaning solids, liquids, and gases from all organisms. It would also subsume all the products of death, including dropped flower petals. *Waste* is a more inclusive word. Consider *Thioploca*. What nourishes its life in the marine sediments? First, a supply of dead organic detritus from the surface—waste. Second, dissolved hydrogen sulfide gas, excreted by the bacteria living in the black muck below—more waste. Third, nitrate ions in the water, most of which have been directly derived from still other bacteria in the water column as they turn the elements within detritus into dissolved ions—still more waste. All of *Thioploca's* diverse nutrients arrive as wastes from other organisms.

A complex biosphere built on excreted products of evolving organisms over billions of years requires us to think beyond dung beetles rolling balls from prairie frisbees. Breathe deeply. Thank the ancient blue-green cyanobacteria that first began splitting water in their quest for scarce hydrogen, so that oxygen was jettisoned as a useless by-product. The global metabolism is now unimaginable without this waste gas, as exhibited in the seasonal rise of the Mauna Loa curve from the global actions of respirers that require oxygen to live. Most of today's photosynthesis comes from plants and algae that themselves need oxygen during internal respiration for building complex molecules. *The waste has become a necessity, even to the organisms that produce it and release it as waste.* Like the climatic effect of the biotic enhancement of weathering, almost all life is bound by the global network built around free oxygen's availability in the air, in the soil, and in the ocean.

Consider the atmosphere's carbon dioxide: More than 99 percent of the entire reservoir has recently been ejected by a respirer rather than a volcano. Nitrogen: More than 99 percent has been discharged from denitrifiers rather than volcanoes. Methane and many other trace gases: More than 99 percent has been expelled from prokaryotes rather than volcanoes. The atmosphere is one giant waste dump. Now, breathe deeply again.

The soil is a waste dump, too. What makes soil soil is the organic matrix from dead life and life's organic and inorganic excreta. Don't neglect the ocean. All life below the surface (except for some chemosynthesizers around the vents) feeds on organic waste. The phosphate and nitrate nutrients, and many other ions, are also wastes. Come to think of it, the only way to get a chemical that is not waste is to grab it before someone is willing to let it go—hence herbivory and predation.

The cycles within Gaia work only because all these waste dumps are simultaneously feedstocks. What makes so much of the gaian matrixes so eminently recyclable? As we saw in Chapter 5, many metabolic pathways can run in reverse. The evolution of the intertwined, cyclic complexity of the global metabolism was possible because many of the action molecules used for one direction in a metabolic pathway can also be used by another organism in reverse. Recall the lament of Lady Macbeth, "What's done cannot be undone." That is true only in the politically charged air of conspirers who are all respirers. But pair such conspirers with photosynthesizers, or nitrogen fixers with denitrifiers, and what is done can usually be undone—indeed *must* be undone to ensure the continuity of Gaia in a bounded system. Stockpiles of cellulose must be raided. Toxins must eventually be detoxified.

Energy is a crucial contributor to Gaia's continuity. Although the material waste of one organism is potential food to another, the sum of all wastes eliminated by any organism will contain less chemical energy than the sum of all forms of energy taken in. Without a renewed energy supply from outside Gaia, the cycles would wind down. In the earliest life, around the vents, the biosphere would have been supplied by the chemical potentials among rock minerals, vent gases, and ions in the water—in other words, by energy from Vulcan. Today the overwhelming source of outside energy is Helios. Photosynthesizers embody energy at a continuous, average rate year round of 150 trillion watts. This is a pittance compared to the total solar flux to Earth, but plenty to power the unique chemistry of life and to flood the gaian matrixes with life's products, giving rise during the course of evolution to the amazing variety of chemical combinations that occur from the air down to the deepest anoxic sediments.

Reversibility of pathways and the inputs of energy from photosynthesizers are principles that have led to a coordinated biota whose members share the gaian matrixes. The complexity woven from free, excreted material by-products and from the climatic side effects links everybody within Gaia into a giant network. Any production from a biochemical guild that affects the general environment acts to coordinate all life. Gaia is thus a system of coordinated evolving systems (the organisms). Note the term is *coordinated*, not *cooperative* or *competitive*. In competing, life gets coordinated. We should jettison that dead-end binary of cooperation and competition in thinking about Gaia. Also, the terminology should not be modified as *optimally coordinated* or *perfectly coordinated*. *Coordinated* by itself is strong enough. Here's the image to keep in mind: Some biochemical guild puts out a substance; then, by relating to that substance, by adapting to its presence, all organisms relate to one another.

This coordination can be seen in the recovery following the giant impact that wiped out the dinosaurs and ninety percent of all species down to the microscopic, at the so-called Cretaceous-Tertiary boundary 65 million years ago. The planet's systems survived. The blow was massive but not fatal, probably because microbes run the show at the base. They are difficult to eliminate (ask any hospital). But ocean productivity did temporarily drop dramatically after the extinction. This drop is apparent in the rock record, not just in fossil abundances but also in life's imprint on carbon isotopes. Chemical conditions took several hundred thousand years to return to normal.

Why did the recovery take so long? Was there a flush of toxic metals from the continents? Did intricate food webs have to re-evolve in the surface ocean to revive productivity fully? Perhaps we should ask what made the recovery so quick. The science of recovery from impacts is so new that we barely have the basic mental tools for proper inquiry. What is the sequence of life's resilience? What, if anything, in the system changed? Did impacts just create new morphologies in the aftermath, known by the fossil transitions, or did the planet's biochemistry shift as well? Future work along these lines, involving chemical changes after mass extinctions and the rise of new biotas, will be at the forefront of gaian science in years to come. In addition to the family of influences on Gaia—Vulcan, Helios, Fortuna—we need to take into account the factor incisively emphasized by my colleague at New York University, geologist Michael Rampino: the powers of destruction and renewal by the Hindu god Shiva, in other words, the cosmos.

Part of the key to gaian resilience after blows from Shiva would derive from the structural principle introduced in Chapter 2: The gaian web of living and nonliving is nearly closed to material transfer across its borders. This is apparent in the high cycling ratios of crucial elements. The holarchy is relatively sealed at the level of Gaia, which is largely why we can speak about a gaian system. I earlier compared this closure to the frame of a painting, bounding an area within which life creates. But such passivity can be an active force when contrasted to a hypothetical situation. Think how different the planet would be if Gaia's ventral border with Vulcan were more permeable. Not only would much larger inputs of gases and minerals come flooding out from Earth's interior, but the ocean, and the atmosphere dissolved in the ocean, would re-enter Vulcan more rapidly. Taking this reasoning to its limit, what if the atmosphere were consumed by Vulcan every few years and replaced with interior gases? Some of these gases, to be sure, would be recycled surface materials, as we saw with the geochemical cycle of carbon. The makeup of a world in which the constituents of Gaia were rapidly exchanged with Vulcan would be difficult to predict. But clearly it would be very different. Certainly life would have less opportunity to influence other life via the shared giant waste bins of the gaian matrixes—waste bins that become feedstocks during evolution.

The paucity of active fluxes across Gaia's borders means that wastes remain in the gaian matrixes long enough for life to strongly affect the concentrations. Closure helps interlace all of life into the network of dependency that makes it continuous across the sweep of time and vulnerable to, but also capable of recovery from, lightning-quick hits from impacts. Closure allows life-mediated pools to accumulate. Gaia's closure gives each form of life time to take advantage of the chemical secretions of the others. This closure yields a kind of inward causation, with the principle that gaian influences are free by-products that forge the physiology of Earth. The resulting high cycling ratios provide a dependable womb for evolution, a cauldron of life and crucial elements. Without the high cycling ratios, human life might have been improbable. Furthermore—who knows?—without the high cycling ratios, life might not have been dense enough for the kinds of biotic interactions necessary for the complex evolution that occurred. I'll leave the question open: Was the stable, predictable, womb-like quality of Gaia necessary for the evolution of consciousness?

⟋ THE NEWEST BIOCHEMICAL GUILD

These days breathing is a special treat for me. This is not to deny that breathing has been a precious activity ever since birth. But an event last winter has since made me far more conscious of the life-giving air.

While writing the middle chapters of this book I went through a Kafkaesque "illness." It began in mid-November with a thumb tip going numb, followed by weeks of electrical zings through my right arm. Then painful cramps in my hands started waking me at night. Perhaps a pinched nerve and carpal tunnel syndrome? No, because soon my toes were also cramping, and half my face lost sensitivity. Symptoms came and went, adding to the confusion and making it tempting to delay seeking treatment, especially when we were snowed in for days at a time in our mountain trailer, an hour and a half drive away from town, where a neurologist arrives once a week by plane, weather permitting.

A doctor at the emergency ward suggested we search for carbon monoxide. I bought a detector and found the culprit: an old wall-mounted propane oven, pressed into service only a few months earlier. Its emissions were not enough to knock us out but were plenty high enough to cause a persistent and progressive poisoning. Thinking all would now be well, I tried my best, as did the neurologist, to assume that relapses were just part of the slow recovery. But these relapses were most unsettling: my arms felt like those of a robot in need of a tune-up, my legs were wobbly, I could not write by hand, and I had long periods of confused, repetitive thoughts (such as spending hours in bed thinking about slicing seeds from an apple). Most worrisome of all were rapid-fire pulsations in my chest, not from my heart, but as though my heart were pounding upon a drum of nerves that resonated at high frequency throughout my body.

Finally, I noticed a pattern in the week-long relapses: They began the evening or morning after trips to town, every couple of weeks. These relapses were especially massive when I had driven alone. I preferred keeping the windows shut and the car toasty warm, whereas when Connie went along, after fifteen minutes in the car she insisted on opening the windows. She felt drowsy—it was early morning, so we rationalized that our hurry to rise had deprived her of sleep. Also, during the drive home she always developed a headache—and she does not get headaches. Hindsight! Finally seeing the pattern, I put the high-tech CO meter into the cab of our old car. Sure enough, the fan pulled in exhaust, which was later found to be leaking from the engine's ancient manifold gasket. All the time we had been dosing ourselves with a second source of carbon monoxide. In wry moments I say I have lived in a Kafka story, taunted by unoxidized carbon atoms.

Ordinary air now feels marvelous. I know that in every breath—especially here in the wilds—there lurks less than half a part per million of my nemesis. The maximum allowable concentration of CO over a stretch of eight hours is fifty ppm. A couple thousand parts per million will kill you within an hour (that is only 0.2 percent in such contaminated air). Our species evolved in an atmosphere with only the tracest of trace amounts of carbon monoxide. Not only are we unable to tolerate air in which carbon monoxide is very much above zero, but nature has also failed to equip us with senses to warn us of the danger. We do not gasp or lunge for the door. Our carbon monoxide naiveté confirms many millions of years of atmospheric makeup. We and atmosphere are one.

Healing from any illness is an opportunity to marvel at the body's ability to restore its complex, dynamic equilibrium, an ability that was built by evolution. While recovering, I could sense millions—really billions—of years right at hand; healing connects the present self to the biological deep past. Time must be thought of not merely as the evolution from cell to complex body, but also as evolution within the larger entity. Gaia is and always was the context for life. Changes in the envelope occurred, to be sure. The most ancient atmosphere would swiftly kill all of us obligate aerobes. But Gaia has always been stable enough to allow the stream of life to carry on in its immense journey.

Rugged nature presents its own dangers. This winter I could have stumbled on a solo hike and frozen to death before rescuers came. Instead, my lesson was to come to the mountains of New Mexico, to live and work in the fresh air—and suffer from technology. As agents of chemical transformation, we humans put wastes into the environment not just from our bodies but also, and far more so, from technologies. We create micro-environments very different from the natural atmosphere in which most of our evolution took place. In Phoenix the tunnels are monitored for carbon monoxide, so motorists can be warned to detour when levels climb dangerously high. But exotic monitors are not needed to detect many of the unusual emissions now bathing parts of the planet. Just take a whiff in a city. Do you smell the newest biochemical guild?

Enough negative statements. The litany of horrors is all too familiar. But as I said in the preface to this book, most of us desire a civilization that includes MRI scans and well-stocked supermarkets. That our actions seem schizophrenic is old news; our productions can debilitate as well as nourish, kill as well as heal. To blame or praise human actions is not my point. Rather, in these few final paragraphs, I hope to convey a nonjudgmental sense of ourselves as Gaia's newest guild.

Our status derives not from bodily exports but from what we do with energy and our technologies. Several thousand years ago, Mediterranean peoples stoked fires to smelt ores. The metals emitted into the air can today be traced in the ice cores of Greenland. Our reach is immense. Altogether new substances are also the mark of this guild: chlorofluorocarbons, for example, and plastics. Speeding up chemical reactions is another of our talents. We oxidize fossil carbon at a rate a hundred times nature's, and thus we perturb the atmosphere's carbon dioxide. As with tree roots, activities can call forth a guild if they broadly and deeply affect the chemistry of Gaia. Agriculture and land use changes wrought by humans are estimated to have increased sediment loads of rivers by about a factor of two worldwide. That's a big impact for a single species.

A call for caution is in order. But let's celebrate, too. Think about our industries for nitrogen fixation. By overcoming the nutrient limitations of natural soils and boosting food productivity, we are actually conserving hinterland that would otherwise be cleared and allowing farms in New England, for example, to return to woods. By itself, our nitrogen fixation is nothing new. Yet consider: all other eukaryotes depend upon prokaryotes for nitrogen fixation, either directly by symbiosis or indirectly by what becomes available in soil and ocean. We are the first eukaryotes to fix nitrogen without the assistance of the ancient ones. That is impressive. True, we cannot yet perform the transformation at room temperature. But perhaps we'll master that trick someday.

We make many new substances and speed up many chemical transformations. What should our guild be called? The chlorofluorocarbon makers? The fossil fuel burners? The sediment loaders? The eukaryotic nitrogen fixers? These are just a few of the possible names. Let's approach the question from another angle.

We are the guild that knows all other guilds.

Here we emphasize not our activities but our knowledge. *Homo sapiens* can come to know all the other guilds. We probe the ocean vents and study exotic metabolisms. We measure the biomass of algae within Antarctic sea ice. We launch satellites to monitor ozone in the stratosphere and productivity on land and across the sea. We pull up sediment samples from lakes to learn, from ancient pollen, which trees lived in the surrounds during the last Ice Age. We bring microbial mats into laboratories to decipher the exchange of gases between members. Our ability to unravel the mysteries of life in the gaian matrixes is cause for celebration. By seeking understanding across the entire planet, we internalize the global holarchy within our minds. In a way, Earth begins to live vibrantly in a conceptual reality, too—a transformation of pattern from matter to mind.

One word to describe this transformation was initially used more than half a century ago by Vladimir Vernadsky and Teilhard de Chardin: *noosphere*. The noosphere is the sphere of the mind. This word pro-claims not just a new biochemical guild but a new kind of guild: the mental guild. I'll leave open the question of whether the noosphere pre-ceded humans.

By our mental models we become conscious of our own activities and those of the whole Earth and beyond. We do not have to wait for the slow meandering of evolution to adapt us to the altered climate and atmospheric chemistry our guild is now creating. Through learn-ing, nitrogen fertilizer can be more scientifically applied—at the best times and proper doses—to limit the run-off that would otherwise en-ter rivers and change estuarine ecosystems. A treaty to phase out the release of chlorofluorocarbons has been concluded, and concerns about slowing the release of carbon dioxide are growing. Recently the Colorado River was allowed, for the first time since its damming, to surge in simulated floods in an attempt to restore the natural cycles of its banks.

Our knowledge *can* take stock of its own ignorance. We become cognizant of what we don't know. Thus in conservation circles we speak of what has come to be called the precautionary principle: Err on the side of caution. Don't do something that may have far-reaching conse-quences without being pretty sure what those consequences will be. The precautionary principle recognizes that our actions affect all other life and thus come back to us. Gaia is a nearly closed system, and we have a long way to go in understanding the global physiology.

I will sign off soon, giving my mind a much-needed break, while I step outside to a glorious sensuous awareness of the connections among the vast atmosphere, the new spring leaves, the depths of evolutionary time, and my own humble breath. Inhale. Exhale. Through this breath I am connected to the deep ocean, the burial of detritus, the cycles kept spinning from year to year. I am connected to the yellow lichen on the tan bluff and floating formaniferans on the other side of the world. I am connected to *Thioploca*, a corn plant, a nematode; to nitrogen-fixing nod-ules, the tundra, the stratosphere. Gratitude is heaven, and heaven is surely here on Earth. Thus I say, thank you, Gaia.

"You're welcome."

"What? Who said that?" The voice sounds like Margaret Mead, self-assured and a bit husky.

"Not Mead. Better attend more carefully."

Now the voice sounds like Raquel Welch, as the demure but technically proficient surgeon's helper in *Fantastic Voyage*. With these subtle switches to embrace multifarious personas, could it be Gaia?

"Could be."

"Wow!"

"Save the fanfare. [sounding now like Whoopi Goldberg] I see you're writing a book about me. Most of it is pretty decent, but such hubris! Humans as the newest biochemical guild and all. Really! From my perspective you humans are overly fecund and very hyperactive respirers, but really nothing special."

"But we burn fossil fuels. That's a first. And we're fixing nitrogen without prokaryotes. We're a new guild."

"You couldn't digest your lunch without my microbes. And your precious fossil fuel business is not at all sustainable on the scale of time that I am accustomed to. Just a flash in the pan. And when fossil fuels are gone, how much nitrogen do you think you'll be able to fix, hmm? Tell me, if you're so smart, what will happen to your cherished New Mexican monsoon with the pending climate change?"

"Uh, I don't know."

"So, you're a Dr. Don't-Know, eh? What about this—are the soil bacteria respiring with more vigor in recent decades?"

"You've caught me blank again, Gaia. But this conversation is beginning to give me a sense of déja vu. Yahweh talked to me like this when I began the book. Do you know him?"

"Him?"

"Tell me more."

"You're starting to annoy me. I didn't come here to talk about me—but you. I shall be brief. You humans need a self-image, a sense of what role you can play in the grand scheme of things. I do rather like your idea of knowledge as the defining quality of your kind, pri-

251

marily because in that way lies hope. But let it be known that you have a heck of a long way to go. Consider your famed Industrial Revolution. Your species put the fire to that transformation with no knowledge of the greenhouse effect it would cause. And you are only just beginning to understand mass extinctions—for so long you blamed them all on me! Indeed! You can't imagine what you have yet to even glimpse about my ways."

"It sounds as if you know a lot more than we do. Perhaps you could help me with a problem. Why is oxygen 21 percent?"

"Why? If you're a member of what you call the noodlesphere . . ."

"Gaia, it's *noosphere*."

"Whatever . . . you'll have to figure out oxygen and everything else yourself. After all, you scientists don't have much regard for revelation, do you?"

"You got me there."

"One last thing. If you humans aspire to be the guild that knows all other guilds, then you've got to assume some responsibility—and show some respect—toward me, my matrixes, and all species. All life has 'responsibilities' within the whole—abilities to respond, linking each to all—so responsibility by itself won't be spanking new with you. But if your guild is going to focus on *knowing*, well, that capability brings a unique kind of responsibility."

"What kind, Gaia?"

"Just think of it as the worthiest of all puzzles."

Bibliography

General

Barlow, Connie. 1991. *From Gaia to Selfish Genes: Selected Writings in the Life Sciences*. Cambridge, Mass.: M.I.T. Press. • Inward and outward causation meet.

Berner, Elizabeth Kay, and Robert A. Berner. 1996. *Global Environment: Water, Air, and Geochemical Cycles*. Upper Saddle River, N.J.: Prentice-Hall. • Simply outstanding general source.

Bunyard, Peter (ed.). 1996. *Gaia in Action: Science of the Living Earth*. Edinburgh, Scotland: Floris. • A scrumptious potpourri guaranteed to please.

Butcher, S. S., R. J. Charlson, G. H. Orians, and G. V. Wolfe (eds.). 1992. *Global Biogeochemical Cycles*. San Diego: Academic. • Multi-authored feast.

Joseph, Lawrence E. 1990. *Gaia: The Growth of an Idea*. New York: St. Martin's Press. • Two for one: history and science.

Lovelock, James E. 1988. *The Ages of Gaia: A Biography of Our Living Earth*. New York: Norton. • Gaia in time and insights into Lovelock's mind.

Lovelock, James E. 1979. *Gaia: A New Look at Life on Earth*. Oxford: Oxford University Press. • The one and only!

Margulis, Lynn, and Dorion Sagan. 1995. *What Is Life?* New York: Simon and Schuster. • Best introduction to the spectrum of the living.

Schlesinger, William H. 1997. *Biogeochemistry: An Analysis of Global Change*, (2d ed). San Diego: Academic. • Turn to this one umpteen times.

Schneider, Stephen H., and Penelope J. Boston (eds.). 1991. *Scientists on Gaia*. Cambridge, Mass.: M.I.T. Press. • Delve in as deeply as you dare.

Volk, Tyler. 1995. *Metapatterns Across Space, Time, and Mind*. New York: Columbia University Press. • Tools for visual thinking.

Westbroek, Peter. 1991. *Life as a Geological Force*. New York: Norton. • Connections between the large and the small.

Williams, George Ronald. 1996. *The Molecular Biology of Gaia*, New York: Columbia University Press. • Highly recommended.

Chapter 1. Breathing of the Biosphere

Bateson, Gregory. 1979. *Mind and Nature: A Necessary Unity*. New York: Dutton. • Ecology as a mental phenomenon.

Heimann, Martin (ed.). 1993. *The Global Carbon Cycle*. Berlin: Springer-Verlag. • Carbon everywhere in multitudinous forms.

Keeling, C. D., J. F. S. Chin, and T. P. Whorf. 1996. Increased activity of northern vegetation inferred from atmospheric CO_2 measurements. *Nature*, *382*, 146–149. • Also see Myneni, R. B., *et al.* 1997. Increased plant growth in the northern high latitudes from 1981 to 1991. *Nature*, *386*, 698–702.

Krumbein, W. E., and A. V. Lapo. 1996. Vernadsky's biosphere as a basis of geophysiology. In *Gaia in Action: Science of the Living Earth*, ed. Peter Bunyard, Edinburgh, Scotland: Floris, pp. 115–134.

Lovelock, James E. 1996. The Gaia hypothesis. In *Gaia in Action: Science of the Living Earth*, ed. Peter Bunyard, Edinburgh, Scotland: Floris, pp. 15–33.

Chapter 2. A Global Holarchy

Allen, Timothy F. H., and Thomas W. Hoekstra. 1992. *Toward a Unified Ecology*. New York: Columbia University Press. • Rich.

Barlow, Connie, and Tyler Volk. 1992. Gaia and evolutionary biology. *BioScience*, *42*, 686–693. • A review of directions to go.

Barlow, Connie, and Tyler Volk. 1990. Open living systems in a closed biosphere: A new paradox for the Gaia debate. *BioSystems*, *23*, 371–384. • Gaia's closure compared to that of organisms.

Bronowski, Jacob. 1970. New concepts in the evolution of complexity: Stratified stability and unbounded plans. *Zygon*, *5*, 18–35.

Capone, Douglas G., et al. 1997. *Trichodesmium*, a globally significant marine cyanobacterium. Science, 276, 1221–1229. • Nitrogen fixation.

Fossing, H., *et al.* 1995. Concentration and transport of nitrate by the mat-forming sulfur bacterium *Thioploca*. *Nature*, *374*, 713–715. • Investigations are underway to determine whether *Thioploca* produces nitrogen gas or ammonium ions from the nitrate (J. Gijs Kuenen, personal communication).

Gallardo, Victor A. 1977. Large benthic microbial communities in sulfide biota under Peru–Chile Subsurface Countercurrent. *Nature*, *268*, 331–332.

Kasting, James F. 1997. Habitable zones around low mass stars and the search for extraterrestrial life. *Origins of Life*, 27, 291–307. • Biotic versus abiotic rates of oxygen production.

Klinger, Lee. 1996. Coupling of soils and vegetation in peatland formation. *Arctic and Alpine Research*, 28, 380–387. • Bogs as climax ecosystems that self-stabilize by altering the soil.

Thamdrup, Bo, and Donald E. Canfield. 1996. Pathways of carbon oxidation in continental margin sediments off central Chile. *Limnology and Oceanography*, 41, 1629–1650. • Analysis of the *Thioploca* community of microbes.

Volk, Tyler. 1995. *Metapatterns Across Space, Time, and Mind*. New York: Columbia University Press.

Wilber, Ken. 1995. *Sex, Ecology, and Evolution*. Boston: Shambhala. • Holarchies and the coevolution of micro and macro.

Chapter 3.   Outer Light, Inner Fire

Broecker, Wallace Smith, and Tsung-Hung Peng. 1982. *Tracers in the Sea*. Palisades, N.Y.: Eldigio Press. • Master work on chemical oceanography, with ocean circulation to boot.

Elderfield, H., and A. Schultz. 1996. Mid-ocean ridge hydrothermal fluxes and the chemical composition of the ocean. *Annual Reviews of Earth and Planetary Science*, 24, 191–224.

Hardie, Lawrence A. 1996. Secular variation in seawater chemistry: An explanation for the coupled secular variation in the mineralogies of marine limestones and potash evaporites over the past 600 million years. *Geology*, 24, 279–283. • Cycles of Vulcan affect ocean chemistry.

Henderson-Sellers, Ann, and Peter J. Robinson. 1986. *Contemporary Climatology*. New York: Wiley. • Excellent general text; see it for the ITCZ, the leaky-bucket model of the greenhouse effect, and the distribution of solar radiation.

Holland, Heinrich D., and Ulrich Petersen. 1995. *Living Dangerously: The Earth, its Resources, and the Environment*. Princeton, N.J.: Princeton University Press. • Everything under the sun, and particularly thorough in the workings of Vulcan.

Huggett, Richard John. 1995. *Geoecology: An Evolutionary Approach*. New York: Routledge. • Interaction of the terrestrial spheres: Atmosphere, biosphere (biota), hydrosphere, pedosphere (soil) and toposphere (relief).

Kump, Lee. 1994. Gaia is a Ruminant. Presented at the 1994 Gaia-in-Oxford meeting. • Lee has allowed me to reinterpret his concept, but he reminded me that he "was specifically referring to the accretionary prism in terms of the rumen, and the subduction zone was the throat down into the stomach of Gaia (upper mantle)."

Chapter 4. The Parts of Gaia

Bonnet, Charles. 1754. Recherches sur l'usage des feuilles dans les plantes. (quoted in Bettex, Albert. 1965. *The Discovery of Nature*. New York: Simon and Schuster.)

Butcher, S. S., R. J. Charlson, G. H. Orians, and G. V. Wolfe (eds.). 1992. *Global Biogeochemical Cycles*. San Diego: Academic. • For distribution of the elements in atmosphere, ocean, soil, and biota.

Fuhrman, Jed A., T. D. Sleeter, C. A. Carlson, and L. M. Proctor. 1989. Dominance of bacterial biomass in the Sargasso Sea and its ecological implications. *Marine Ecology Progress Series*, 57, 207–217. • Surface areas of marine bacteria.

Global Change Newsletter. September 1996 (No. 27), p. 9. • Land classification for the International Geosphere-Biosphere Program.

Gould, Stephen Jay. 1996. *Full House: The Spread of Excellence from Plato to Darwin*. New York: Harmony. • Discusses upper estimates for bacterial biomass.

Heaney, Seamus. 1996. *The Spirit Level*. New York: Farrar, Straus & Giroux.

Hunt, E. Raymond, Jr., S. C. Piper, R. Nemani, C. D. Keeling, R. D. Otto, and S. W. Running. 1996. Global net carbon exchange and intra-annual atmospheric CO_2 concentrations predicted by an ecosystem process model and three-dimensional atmospheric transport model. *Global Biogeochemical Cycles*, 10, 431–456. • The BIOME Model.

Kump, Lee R., and James E. Lovelock. 1995. The geophysiology of climate. In *Future Climates of the World: A Modeling Perspective*, ed. A. Henderson-Sellers, Amsterdam: Elsevier, pp. 537–553.

Lovelock, James E., and Lee R. Kump. 1994. Failure of climate regulation in a geophysiological model. *Nature*, 369, 732–734. • Marine algae and land plants as biochemical parts of Gaia.

Mankiewicz, Paul S. 1991. The macromolecular matrix of plant cell walls as a major gaian interfacial regulator in terrestrial environments. In: *Scientists on Gaia*, ed. S. H. Schneider and P. J. Boston, Cambridge, Mass.: M.I.T. Press, pp. 309–319. • Awesome power of living surfaces.

Margulis, Lynn, and Karlene V. Schwartz. 1988. *Five Kingdoms: An Illustrated Guide to the Phyla of Life on Earth.*. New York: Freeman.

Morowitz, Harold J. 1992. *Beginnings of Cellular Life: Metabolism Recapitulates Biogenesis*. New Haven, Conn.: Yale University Press. • Thought-provoking.

Morowitz, Harold J. 1979. *Energy Flow in Biology: Biological Organization as a Problem in Thermal Physics*. Woodbridge, Conn.: Ox Bow Press.

van Breemen, N. 1993. Soils as biotic constructs favoring net primary productivity. *Geoderma*, 57, 183–211. • A three-page comment by Lovelock follows this paper.

Vernadsky, Vladimir I. 1997/1926. *The Biosphere*, ed. Mark McMenamin, New York: Copernicus/Springer-Verlag. • Life as a geological force.

Woese, Carl R., Otto Kandler, and Mark L. Wheelis. 1990. Towards a natural system of organisms: Proposal for the domains Archaea, Bacteria, and Eucarya. *Proceedings of the National Academy of Sciences*, 87, 4576–4579.

Chapter 5. Worldwide Metabolisms

Badger, M.R., and T.J. Andrews. 1987. Co-evolution of Rubisco and CO_2 concentrating mechanisms. In *Progress in Photosynthesis Research*, Vol. 3, ed. J. Biggens, Dordrecht, the Netherlands: Martinus Nijhoff, pp. 601–609.

Caldwell, Mark. 1995. The amazing all-natural light machine. *Discover*, December, 88–95. • Analysis of the light-collecting antenna of a purple sulfur bacterium.

Danks, S. M., E. H. Evans, and P. A. Whittaker. 1983. *Photosynthetic Systems*. New York: Wiley.

Ehleringer, James R., Rowan F. Sage, Lawrence B. Flanagan, and Robert W. Pearcy. 1991. Climate change and the evolution of C_4 photosynthesis. *Trends in Ecology and Evolution*, 6, 95–99.

Falkowski, P. G., and J. A. Raven. 1997. *Aquatic Photosynthesis*. Oxford: Blackwell.

Jeffrey, S. W. 1980. Algal pigment systems. In *Primary Productivity in the Sea*, ed. Paul G. Falkowski, New York: Plenum, pp. 33–58.

Kandler, Otto. 1994. The early diversification of life. In *Early Life on Earth*, Nobel Symposium No. 84, ed. Stefan Bengtson, New York: Columbia University Press, pp. 152–160.

Kozaki, Akiko, and Go Takeba. 1996. Photorespiration protects plants from photooxidation. *Nature*, 384, 557–560.

Marschner, Horst. 1995. *Mineral Nutrition of Higher Plants*. San Diego: Academic. • Siderophores and acid phosphatases.

Miyashita, Hideaki, *et al.* 1996. Chlorophyll-d as a major pigment. *Nature*, 383, 402. • Another type of chlorophyll.

Monson, R. K., and B. D. Moore. 1989. On the significance of C_3-C_4 intermediate photosynthesis to the evolution of C_4 photosynthesis. *Plant, Cell, and Environment*, 12, 689–699.

Pierson, Beverly K. 1994. The emergence, diversification, and role of photosynthetic eubacteria. In *Early Life on Earth*, Nobel Symposium No. 84, ed. Stefan Bengtson, New York: Columbia University Press, pp. 161–180.

Taiz, Lincoln, and Eduardo Zeiger. 1991. *Plant Physiology*. Redwood City, Calif.: Benjamin/Cummings. • Excellent general text.

Wächterhäuser, Günter. 1994. Vitalysts and virulysts: A theory of self-expanding reproduction. In *Early Life on Earth*, Nobel Symposium No. 84, ed. Stefan Bengtson, New York: Columbia University Press, pp. 124–132.

Williams, George Ronald. 1996. *The Molecular Biology of Gaia*, New York: Columbia University Press. • My quotes are from his page 179.

Chapter 6. Embodied Energy

Berner, Elizabeth Kay, and Robert A. Berner. 1996. *Global Environment: Water, Air, and Geochemical Cycles*. Upper Saddle River, N.J.: Prentice-Hall. • Numbers for the terrestrial system of calcium and phosphorus.

Bierce, Ambrose. 1911. *The Devil's Dictionary*. Cleveland, Ohio: World.

Bugbee, Bruce. 1992. Determining the potential productivity of food crops in controlled environments. *Advances in Space Research*, *12(5)*, 85–95. • Energy cascade.

Bugbee, Bruce, and Oscar Monje. 1992. The limits of crop productivity. *BioScience*, *42*, 494–502. • More energy cascade.

Guyton, Arthur C. 1986. *Textbook of Medical Physiology*, (7th ed). Philadelphia: Saunders. • Human composition.

Taiz, Lincoln, and Eduardo Zeiger. 1991. *Plant Physiology*. Redwood City, Calif.: Benjamin/Cummings. • Composition of typical plant tissue; numbers used to compute annual terrestrial uptake of elements.

Volk, Tyler. 1996. Miniaturizing simplified agro-ecosystems for advanced life support. *Ecological Engineering*, *6*, 99–108. • Overview of NASA's CELSS program.

Volk, Tyler, Bruce Bugbee, and Ray M. Wheeler. 1995. An approach to crop modeling with the energy cascade. *Life Support and Biosphere Science*, *1*, 119–127. • Still more energy cascade.

Volk, Tyler, and John D. Rummel. 1989. The case for cellulose production on Mars. In *The Case for Mars III; Strategies for Exploration — Technical*, Vol. 75, American Astronautical Society Science and Technology Series, pp. 87–94.

Wheeler, Raymond M., *et al.* 1993. A data base of crop nutrient use, water use, and carbon dioxide exchange in a 20 square meter growth chamber: Wheat as a case study. *Journal of Plant Nutrition*, *16*, 1881–1915. • Composition of wheat plants.

Chapter 7. The Music of This Sphere

Berner, Elizabeth Kay, and Robert A. Berner. 1996. *Global Environment: Water, Air, and Geochemical Cycles*. Upper Saddle River, N.J.: Prentice-Hall. • Numbers for the marine system of phosphorus and calcium; sulfate reduction.

Charlson, R. J., J. E. Lovelock, M. O. Andreae, and S. G. Warren. 1987. Oceanic phytoplankton, atmospheric sulfur, cloud albedo, and climate. *Nature*, *274*, 246–248. • Classic paper on the climatic effects of DMS.

Coale, Kenneth H., *et al.* 1996. Control of community growth and export production by upwelled iron in the equatorial Pacific Ocean. *Nature, 379*, 612–624.

Coale, Kenneth H., *et al.* 1996. A massive phytoplankton bloom induced by an ecosystem-scale iron fertilization experiment in the equatorial Pacific Ocean. *Nature, 383*, 495–501.

Falkowski, Paul G. 1995. Ironing out what controls primary production in the nutrient rich waters of the open ocean. *Global Change Biology, 1*, 161–163.

Gruber, Nicolas, and Jorge L. Sarmiento. 1997. Global patterns of marine nitrogen fixation and denitrification. *Global Biogeochemical Cycles, 11*, 235–266. • Using ocean chemistry to trace the patterns of biochemical guilds.

Krumbein, Wolfgang E., and Peter K. Swart. 1983. The microbial carbon cycle. In *Microbial Geochemistry*, ed. W. E. Krumbein, Oxford: Blackwell Scientific, pp. 5–62.

Likens, Gene E., and E. Herbert Borman. 1995. *Biogeochemistry of a Forested Ecosystem*, (2d ed). New York: Springer-Verlag. • The Hubbard Brook Experimental Forest.

Rampino, Michael R., and Tyler Volk. 1988. Mass extinctions, atmospheric sulfur and climatic warming at the K/T boundary, *Nature, 332*, 63–65. • Post-impact warming from die-off of DMS-emitters.

Turner, Suzanne M., *et al.* Increased dimethyl sulfide concentrations in sea water from *in situ* iron enrichment. *Nature, 383*, 513–517.

Tyrrell, T., and C. S. Law. 1997. Low nitrate:phosphate ratio in the global ocean. *Nature, 387*, 793–796. • The ratio is very low in 2 percent of the data, indicating sites of denitrification.

Watson, Andrew J. 1997. Volcanic iron, CO_2, ocean productivity, and climate. *Nature, 385*, 587–588. • Connections all the way back to the Ice Age.

Chapter 8.   Gaia in Time

Algeo, Thomas J., Berner, R. A., Maynard, J. B., and Scheckler, S. E. 1995. Late Devonian oceanic anoxic events and biotic crises: "Rooted" in the evolution of vascular land plants? *GSA Today, 5*(3).

Barlow, Connie. 1997. *Green Space, Green Time: The Way of Science*. New York: Copernicus/Springer-Verlag. • Importance of visualizing deep time in the biological present.

Berner, Robert A. 1997. The rise of plants and their effect on weathering and atmospheric CO_2. *Science, 276*, 5 44–546.

Caldeira, Ken. 1989. Evolutionary pressures on planktonic production of atmospheric sulfur. *Nature, 337*, 732–734. • Also see Caldeira, Ken. 1991. Evolutionary pressures on planktonic dimethyl sulfide production. In *Scientists on Gaia*, ed. S. H. Schneider and P. J. Boston, Cambridge, Mass.: M.I.T. Press, pp. 153–158.

Canfield, Donald E., and Andreas Teske. 1996. Late Proterozoic rise in atmospheric oxygen concentration inferred from phylogenetic and sulfur-isotope studies. *Nature*, *382*, 127–132.

Des Marais, David. 1996. Work reported in the *Lunar and Planetary Information Bulletin*, No. 81, p. 8.

Falkowski, Paul G. 1997. Evolution of the nitrogen cycle and its influence on the biological sequestration of CO_2 in the ocean. *Nature*, *387*, 272–275. • He holds that denitrification preceded nitrogen fixation in Earth history.

Graham, Jeffrey B., Robert Dudley, Nancy Aguilar, and Carl Gans. 1995. Implications of the late Paleozoic oxygen pulse for physiology and evolution. *Nature*, *375*, 117–120.

Holland, Heinrich D. 1994. Early Proterozoic atmospheric change. In *Early Life on Earth*, Nobel Symposium No. 84, ed. Stefan Bengtson, New York: Columbia University Press, pp. 237–244. • Oxygen rise at 2 billion years ago.

Kasting, James F., David H. Eggler, and Stuart P. Raeburn. 1993. Mantle redox evolution and the oxidation state of the Archean atmosphere. *Journal of Geology*, *101*, 245–257.

Kasting, James F., Daniel P. Whitmire, and Ray T. Reynolds. 1993. Habitable zones around main sequence stars. *Icarus*, *101*, 108–128. • Best analysis of this question.

Keeling, Ralph F., Stephen C. Piper, and Martin Heimann. 1996. Global and hemispheric CO_2 sinks deduced from changes in atmospheric O_2 concentration. *Nature*, *381*, 218–221. • Oxygen drops as carbon dioxide rises.

Mancinelli, Rocco L., and Christopher P. McKay. 1988. The evolution of nitrogen cycling. *Origins of Life and Evolution of the Biosphere*, *18*, 311–325.

McMenamin, Mark A. S., and Dianna L. S. McMenamin. 1994. *Hypersea: Life on Land*. New York: Columbia University Press.

Nisbet, Euan G. 1995. Archaean ecology: A review of evidence for the early development of bacterial biomes, and speculations on the development of a global-scale biosphere. In *Early Precambrian Processes*, ed. M. P. Coward and A. C. Ries, Geological Society of London Special Publication No. 95, pp. 27–51. • Ideas galore. Also see Nisbet, E. G., and C. M. R. Fowler. 1996. Some liked it hot. *Nature*, *382*, 404–405.

Rampino, Michael R. 1991. Gaia versus Shiva: Cosmic effects on the longterm evolution of the terrestrial biosphere. In *Scientists on Gaia*, ed. S. H. Schneider and P. J. Boston, Cambridge, Mass.: M.I.T. Press, pp. 382–390.

Robinson, Jennifer. 1990. Lignin, land plants, and fungi: Biological evolution affecting Phanerozoic oxygen balance. *Geology*, *15*, 607–610.

Schwartzman, David W., and Tyler Volk. 1991. Biotic enhancement of weathering and surface temperatures on earth since the origin of life. *Palaeogeography, Palaeoclimatology, Palaeoecology* (Global and Planetary Change Section), *90*, 357–371.

Schwartzman, David W., and Tyler Volk. 1989. Biotic enhancement of weathering and the habitability of Earth. *Nature, 340*, 457–460.

Schwartzman, David W., Mark McMenamin, and Tyler Volk. 1993. Did surface temperatures constrain microbial evolution? *BioScience, 43*, 390–393.

Vernadsky, Vladimir I. 1945. The biosphere and the noosphere. *American Scientist, 33*, 1–12.

Volk, Tyler. 1989. Sensitivity of climate and atmospheric CO_2 to deep-ocean and shallow-ocean carbonate burial. *Nature, 337*, 637–640.

Volk, Tyler. 1987. Feedbacks between weathering and atmospheric CO_2 over the last 100 million years. *American Journal of Science, 287*, 763–779.

Index

Outward influences, 49–50
Oxygen
  amount in atmosphere, 3, 5, 222, 252
  detritus and, 35–36
  in Earth history, 139, 145, 224, 227–230
  gas, defined, 110
  generation, 110, 132, 157, 170–171
  lack of in ocean, 35–37, 39, 76, 231
  and nitrogen fixation, 149–151
  and plant growth, 139
Ozone, 4, 169, 174, 224

Parts per million, defined, 7
Phosphate, defined, 202
Phosphorus, 173
  distribution, 116
  higher in deep water, 202–203
  marine cycling ratio, 191–200
  and plant growth, 151
  ratio to carbon in life, 192–193
  terrestrial cycling ratio, 178–182
  See also nitrogen-to-phosphorus ratio
Photosynthesis, 51, 54, 81, 86, 98, 132
  active photons, 160
  amount of global, 58, 174
  amount of marine, 55, 165–166, 192
  amount of terrestrial, 58, 159–166, 192
  C_3 and C_4 pathways, 140–141, 161, 235
  calcium required, 183
  depleting ocean's surface nutrients, 113, 202
  equation, 145
  evolution, 133–136, 143–144, 225–227
  as flux in diagram, 17
  marine versus terrestrial, 197–200
  phosphorus required, 179–182, 194–195, 198
  in seasonal oscillation of CO_2, 9–13, 221–222
  See also carbon dioxide, affecting plant growth; chlorophyll; energy cascade
Physiology, of body, 14, 16, 23–24, 75–78, 112, 124, 190–191
Pine Woods Park, 31–33, 44, 46, 53, 60, 73

Phytoplankton, 33, 165, 192–193, 209, 240
  See also plankton
Plankton, calcareous, 90, 232–233
Plants 41, 95, 119, 126–127, 133, 140, 176–177
  grasses, 96, 142
  lignin, 193, 198, 231
  soy plants, 155, 205
  surface area of roots, 121
  trees, 231–232
  wheat, 17, 142, 155, 205
Pools
  as causal factors, 17, 19, 24
  defined, 16
Ppm, *see* parts per million
Prime directive for gaian inquiry, 27–29, 42, 45, 105
  binary in, 27
  See also Earth, without life or some forms of life
Prokaryotic cells, 225
  defined, 97
Protoctists, introduced, 96–97

Rampino, Michael, 244
Reservoirs, *see* pools
Respiration, 19, 52, 54, 103, 144–147, 243
  equation, 145
  as reversal of synthesis pathway, 146
  in seasonal oscillation of CO_2, 9–13
Responsibility, 252
Richardson, Lewis, 71
Rivers, 118, 179, 183, 189–191, 192, 206, 248
Robinson, Jennifer, 231
Roots
  surface area, 121
  in weathering, 232, 237
Rubisco, 136–144, 147, 153, 156, 161, 193, 226
  defined, 137
RuBP, defined, 137

Schwartzman, David, 227, 230, 236, 238–239